# THE PASSING OF THE GREAT RACE

## THE RACIAL BASIS OF EUROPEAN HISTORY

MADISON GRANT

PAST
CHAIRMAN, NEW YORK ZOOLOGICAL SOCIETY;
TRUSTEE, AMERICAN MUSEUM OF NATURAL HISTORY;
COUNCILOR, AMERICAN GEOGRAPHICAL SOCIETY

Originally Published in 1916

**Note:** This edition is based on Grant's original, 1916, publication. It is thus free of later alterations. Grant's conclusions are not endorsed in any way, whatsoever. Rather, it is presented with an eye towards understanding history as it actually unfolded, blemishes and all.

Charts and maps are exactly as produced by Grant. Maps are in their original coloring. Charts had their black and white balances adjusted for easier reading, but are otherwise also as originally published in 1916.

*The Passing of the Great Race: The Racial Basis of European History*
By Madison Grant

ISBN: 978-1-947844-99-5



**Note:** This edition is based on Grant's original, 1916, publication. It is thus free of later alterations. Grant's conclusions are not endorsed in any way, whatsoever. Rather, it is presented with an eye towards understanding history as it actually unfolded, blemishes and all.

Charts and maps are exactly as produced by Grant. Maps are in their original coloring. Charts had their black and white balances adjusted for easier reading, but are otherwise also as originally published in 1916.

# TABLE OF CONTENTS

## CHARTS AND MAPS

### CHARTS



### MAPS
### AT THE END OF VOLUME

*(After the pages included for notes)*

# PREFACE

European history has been written in terms of nationality and of language, but never before in terms of race; yet race has played a far larger part than either language or nationality in moulding the destinies of men; race implies heredity, and heredity implies all the moral, social, and intellectual characteristics and traits which are the springs of politics and government. Quite independently and unconsciously the author, never before a historian, has turned this historical sketch into the current of a great biological movement, which may be traced back to the teachings of Galton and Weismann, beginning in the latter third of the nineteenth century. This movement has compelled us to recognize the superior force and stability of heredity, as being more enduring and potent than environment. This movement is also a reaction from the teachings of Henri Taine among historians and of Herbert Spencer among biologists, because it proves that environment and, in the case of man, education have an immediate, apparent, and temporary influence, while heredity has a deep, subtle, and permanent influence on the actions of men.

Thus the racial history of Europe, which forms the author's main outline and subject and which is wholly original in treatment, might be paraphrased as the heredity history of Europe. It is history as influenced by the hereditary impulses, predispositions, and tendencies which as highly distinctive racial traits date back many thousands of years and were originally formed when man was still in the tribal state, long before the advent of civilization.

In the author's opening chapters these traits and tendencies are commented upon as they are observed to-day under the varying influences of migration and changes of social and physical environment. In the chapters relating to the racial history of Europe we enter a new and fascinating field of study, which I trust the author himself may some day expand into a longer story. There is no gainsaying that this is the correct scientific method of approaching the problem of the past.

The moral tendency of the heredity interpretation of history is for our day and generation, and is in strong accord with the true spirit of the modern eugenics movement in relation to patriotism, namely, the conservation and multiplication for our country of the best spiritual, moral, intellectual, and physical forces of heredity; thus only will the integrity of our institutions be maintained in the future. These divine forces are more or less sporadically distributed in all races, some of them are found in what we call the lowest races, some are scattered widely throughout humanity, but they are certainly more widely and uniformly distributed in some races than in others.

Thus conservation of that race, which has given us the true spirit of Americanism, is not a matter either of racial pride or of racial prejudice; it is a matter of love of country, of a true sentiment which is based upon knowledge and the lessons of history, rather than upon the sentimentalism which is fostered by ignorance. If I were asked: What is the greatest danger which threatens the American republic to-day? I would certainly reply: The gradual dying out among our people of those hereditary traits through which the principles of our religious, political, and social foundations were laid down, and their insidious replacement by traits of less noble character.

Henry Fairfield Osborn.
July 13, 1916.

# INTRODUCTION

The following pages are devoted to an attempt to elucidate the meaning of history in terms of race; that is, by the physical and psychical characters of the inhabitants of Europe instead of by their political grouping, or by their spoken language. Practically all historians, while using the word race, have relied on tribal or national names as its sole definition. The ancients, like the moderns, in determining ethnical origin, did not look beyond a man's name, language, or country, and the actual information furnished by classic literature on the subject of physical characters is limited to a few scattered and often obscure remarks.

Modern anthropology has demonstrated that racial fines are not only absolutely independent of both national and linguistic groupings, but that in many cases these racial fines cut through them at sharp angles and correspond closely with the divisions of social cleavage. The great lesson of the science of race is the immutability of somatological or bodily characters, with which is closely associated the immutability of psychical predispositions and impulses. This continuity of inheritance has a most important bearing on the theory of democracy and still more upon that of socialism, and those engaged in social uplift and in revolutionary movements are consequently usually very intolerant of the limitations imposed by heredity.

Democratic theories of government in their modem form are based on dogmas of equality formulated some hundred and fifty years ago, and rest upon the assumption that environment and not heredity is the controlling factor in human development. Philanthropy and noble purpose dictated the doctrine expressed in the Declaration of Independence, the document which to-day constitutes the actual basis of American institutions. The men who wrote the words, "we hold these truths to be self-evident, that all men are created equal," were themselves the owners of slaves, and despised Indians as something less than human. Equality in their minds meant merely that they were just as good Englishmen as their brothers across the sea. The words "that all men are created equal" have since been subtly falsified by adding the word "free," although no such expression is found in the original document, and the teachings based on these altered words in the American public schools of to-day would startle and amaze the men who formulated the Declaration.

The laws of nature operate with the same relentless and unchanging force in human affairs as in the phenomena of inanimate nature, and the basis of the government of man is now and always has been, and always will be, force and not sentiment, a truth demonstrated anew by the present world conflagration.

It will be necessary for the reader to strip his mind of all preconceptions as to race, since modern anthropology, when applied to history, involves an entire change of definition. We must, first of all, realize that race pure and simple, the physical and psychical structure of man, is something entirely distinct from either nationality or language, and that race lies to-day at the base of all the phenomena of modem society, just as it has done throughout the unrecorded eons of the past.

The antiquity of existing European populations, viewed in the fight thrown upon their origins by the discovery of the last few decades, enables us to carry back history and prehistory into periods so remote that the classic world is but of yesterday. The living peoples of Europe consist of layer after layer of diverse racial elements in varying proportions, and historians and anthropologists, while studying these populations, have been concerned chiefly with the recent strata, and have neglected the more ancient and submerged types.

Aboriginal populations from time immemorial have been again and again swamped under floods of newcomers and have disappeared for a time from historic view. In the course of centuries, however, these primitive elements have slowly reasserted their physical type and have gradually bred out their conquerors, so that the racial history of Europe has been in the past, and is to-day a story of the repression and resurgence of ancient races.

Invasions of new races have ordinarily arrived in successive waves, the earlier ones being quickly absorbed by the conquered, while the later arrivals usually maintain longer the purity of their type. Consequently the more recent elements are found in a less mixed state than the older, and the more primitive strata of the population always contain physical traits derived from still more ancient predecessors.

Man has inhabited Europe in some form or other for hundreds of thousands of years, and during all this lapse of time the population has been as dense as the food supply permitted. Tribes in the hunting stage are necessarily of small size, no matter how abundant the game, and in the Paleolithic period man probably existed only in specially favorable localities, and in relatively small communities.

In the Neolithic and Bronze periods domesticated animals and the knowledge of agriculture, although of primitive character, afforded an enlarged food supply, and the population in consequence greatly increased. The lake dwellers of the Neolithic were, for example, relatively numerous. With the clearing of the forests arid the draining of the swamps during the Middle Ages and, above all, with the industrial expansion of the last century, the population multiplied with great rapidity. We can, of course, form little or no estimate of the numbers of the Paleolithic population of Europe, and not much more of those of Neolithic times, but even the latter must have been very small in comparison with the census of to-day.

Some conception of the growth of population in recent times may be based on the increase in England. It has been computed that Saxon England at the time of the Conquest contained about 1,500,000 inhabitants; at the time of Queen Elizabeth the population was about 4,000,000, while in 1911 the census gave for the same area some 35,000,000.

The immense range of the subject of race in connection with history from its nebulous dawn, and the limitations of space, require that generalizations must often be stated without mention of exceptions. These sweeping statements may even appear to be too bold, but they rest, to the best of the writer's belief, upon solid foundations of facts, or else are legitimate conclusions from evidence now in hand. In a science as recent as modern anthropology, new facts are constantly revealed and require the modification of existing hypotheses. The more the subject is studied the more provisional even the best-sustained theory appears, but modern research opens a vista of vast interest and significance to man, now that we have discarded the shackles of former false view-points and are able to discern, even though dimly, the solution of many of the problems of race. New data will in the future inevitably expand, and perhaps change our ideas, but such facts as are now in hand, and the conclusions based thereupon, are provisionally set forth in the following chapters, and necessarily often in a dogmatic form.

The statements relating to time have presented the greatest difficulty, as the authorities differ widely, but the dates have been fixed with extreme conservatism and the writer believes that whatever changes in them are hereafter required by further investigation and study, will result in pushing them back and not forward in prehistory.

The dates given in the chapter of "Paleolithic Man" are frankly taken from the most recent authority on this subject, "The Men of the Old Stone Age," by Professor Henry Fairfield Osborn, and the writer desires to take this opportunity to acknowledge his great indebtedness to this source of information, as well as to Mr. M. Taylor Pyne and to Mr.

Charles Stewart Davison for their assistance and many helpful suggestions.

The author also wishes to acknowledge a debt of gratitude to Professor William Z. Ripley's great work on "The Races of Europe," which contains a vast array of anthropological data, maps, and type portraits, providing a mine of information upon which the author has drawn freely, for the present distribution of the three primary races of Europe.

The American Geographical Society and its staff, particularly Mr. Leon Dominian, have also been of great assistance in the preparation of the maps contained herein, and this occasion is taken by the writer to express his deep appreciation for their assistance.

To
MY FATHER

# THE PASSING OF THE GREAT RACE

# PART I

## RACE, LANGUAGE, AND NATIONALITY

## I
## RACE AND DEMOCRACY

Failure to recognize the clear distinction between race and nationality and the still greater distinction between race and language, the easy assumption that the one is indicative of the other, has been in the past a serious impediment to an understanding of racial values. Historians and philologists have approached the subject from the view-point of linguistics, and as a result we have been burdened with a group of mythical races, such as the Latin, the Aryan, the Caucasian, and, perhaps, most inconsistent of all, the "Celtic" race.

Man is an animal differing from his fellow inhabitants of the globe, not in kind but only in degree of development, and an intelligent study of the human species must be preceded by an extended knowledge of other mammals, especially the primates. Instead of such essential training, anthropologists often seek to qualify by research in linguistics, religion, or marriage customs, or in designs of pottery or blanket weaving, all of which relate to ethnology alone.

The question of race has been further complicated by the effort of old-fashioned theologians to cramp all mankind into the scant six thousand years of Hebrew chronology, as expounded by Archbishop Ussher. Religious teachers have also maintained the proposition not only that man is something fundamentally distinct from other living creatures, but that there are no inherited differences in humanity that cannot be obliterated by education and environment.

It is, therefore, necessary at the outset for the reader to thoroughly appreciate that race, language, and nationality are three separate and distinct things, and that in Europe these three elements are only occasionally found persisting in combination, as in the Scandinavian nations.

To realize the transitory nature of political boundaries, one has only to consider the changes of the past century, to say nothing of those which may occur at the end of the present war. As to language, here in America we daily hear the English language spoken by many men who possess not one drop of English blood, and who, a few years since, knew not one word of Saxon speech.

As a result of certain religious and social doctrines, now happily becoming obsolete, race consciousness has been greatly impaired among civilized nations, but in the beginning all differences of class, of caste, and of color, marked actual lines of race cleavage.

In many countries the existing classes represent races that were once distinct. In the city of New York, and elsewhere in the United States, there is a native American aristocracy resting upon layer after layer of immigrants of lower races, and the native American, while, of course, disclaiming the distinction of a patrician class, nevertheless has, up to this time, supplied the leaders of thought and the control of capital, of education, and of the religious ideals and altruistic bias of the community.

In the democratic forms of government the operation of universal suffrage tends toward the selection of the average man for public office rather than the man qualified by birth, education, and integrity. How this scheme of administration will ultimately work out remains to be seen, but from a racial point of view, it will inevitably increase the preponderance of the lower types and cause a corresponding loss of efficiency in the community as a whole.

The tendency in a democracy is toward a standardization of type and a diminution of the influence of genius. A majority must of necessity be inferior to a picked minority, and it always resents specializations in which it cannot share. In the French Revolution the majority, calling itself "the people," deliberately endeavored to destroy the higher type, and something of the same sort was, in a measure, done after the American Revolution by the expulsion of the Loyalists and the confiscation of their lands.

In America we have nearly succeeded in destroying the privilege of birth; that is, the intellectual and moral advantage a man of good stock brings into the world with him. We are now engaged in destroying the privilege of wealth; that is, the reward of successful intelligence and industry, and in some quarters there is developing a tendency to attack the privilege of intellect and to deprive a man of the advantages of an early and thorough education. Simplified spelling is a step in this direction. Ignorance of English grammar or classic learning must not be held up as a reproach to the political and social aspirant.

Mankind emerged from savagery and barbarism under the leadership of selected individuals whose personal prowess, capacity, or wisdom gave them the right to lead and the power to compel obedience. Such leaders have always been a minute fraction of the whole, but as long as the tradition of their predominance persisted they were able to use the brute strength of the unthinking herd as part of their own force, and were able to direct at will the blind dynamic impulse of the slaves, peasants, or lower classes. Such a despot had an enormous power at his disposal which, if he were benevolent or even intelligent, could be used, and most frequently was used, for the general uplift of the race. Even those rulers who most abused this power put down with merciless rigor the antisocial elements, such as pirates, brigands, or anarchists, which impair the progress of a community, as disease or wounds cripple an individual.

True aristocracy is government by the wisest and best, always a small minority in any population. Human society is like a serpent dragging its long body on the ground, but with the head always thrust a little in advance and a little elevated above the earth. The serpent's tail, in human society represented by the antisocial forces, was in the past dragged by sheer force along the path of progress. Such has been the organization of mankind from the beginning, and such it still is in older communities than ours. What progress humanity can make under the control of universal suffrage, or the rule of the average, may find a further analogy in the habits of certain snakes which wiggle sideways and disregard the head with its brains and eyes. Such serpents, however, are not noted for their ability to make rapid progress.

To use another simile, in an aristocratic as distinguished from a plutocratic, or democratic organization, the intellectual and talented classes form the point of the lance, while the massive shaft represents the body of the population and adds by its bulk and weight to the penetrative impact of the tip. In a democratic system this concentrated force at the top is dispersed throughout the mass, supplying, to be sure, a certain amount of leaven, but in the long run the force and genius of the small minority is dissipated, if not wholly lost. Vox populi, so far from being Vox Dei, thus becomes an unending wail for rights, and never a chant of duty.

Where a conquering race is imposed on another race the institution of slavery often arises to compel the servient race to work, and to introduce it forcibly to a higher form of civilization. As soon as men can be induced to labor to supply their own needs slavery becomes wasteful and tends to vanish. Slaves are often more fortunate than freemen when treated with reasonable humanity, and when their elemental wants of food, clothing, and shelter are supplied.

The Indians around the fur posts in northern Canada were formerly the virtual bond slaves of the Hudson Bay Company, each Indian and his squaw and pappoose being adequately supplied with simple food and equipment. He was protected as well against the white man's rum as the red man's scalping parties, and in return gave the Company all his peltries—the whole product of his year's work. From an Indian's point of view this was nearly an ideal condition, but was to all intents serfdom or slavery. When, through the opening up of the country, the continuance of such an archaic system became an impossibility, the Indian sold his furs to the highest bidder, received a large price in cash, and then wasted the proceeds in trinkets instead of blankets, and in mm instead of flour, with the result that he is now gloriously free, but is on the highroad to becoming a diseased outcast. In this case of the Hudson Bay Indian the advantages of the upward step from serfdom to freedom are not altogether clear. A very similar condition of vassalage existed until recently among the peons of Mexico, but without the compensation of an intelligent and provident ruling class.

In the same way serfdom in mediæval Europe apparently was a device through which the landowners overcame the nomadic instincts of their tenantry. Years are required to bring land to its highest productivity, and agriculture cannot be successfully practised even in well-watered and fertile districts by farmers who continually drift from one locality to another. The serf or villein was, therefore, tied by law to the land, and could not leave except with his master's consent. As soon as these nomadic instincts ceased to exist serfdom vanished. One has only to read the severe laws against vagrancy in England, just before the Reformation, to realize how widespread and serious was this nomadic instinct.

Here in America we have not yet forgotten the wandering instincts of our Western pioneers, which in that case proved to be beneficial to every one except the migrants.

# PART II

## THE PHYSICAL BASIS OF RACE

In the modern and scientific study of race we have long discarded the Adamic theory that man is descended from a single pair, created a few thousand years ago in a mythical Garden of Eden somewhere in Asia, to spread later over the earth in successive waves.

Many of the races of Europe, both living and extinct, did come from the East through Asia Minor or by way of the African littoral, but most of the direct ancestors of existing populations have inhabited Europe for many thousands of years. During that time numerous races of men have passed over the scene. Some undoubtedly have utterly vanished, and some have left their blood behind them in the Europeans of to-day.

It is a fact, however, that Asia was the chief area of evolution and differentiation of man, and that the various groups had their main development there, and not on the peninsula we call Europe.

We now know, since the elaboration of the Mendelian Laws of Inheritance, that certain bodily characters, the so-called unit characters, such as skull shape, stature, eye color, hair color, and nose form, are transmitted in accordance with fixed mathematical laws, and, further, that various unit characters which are normally correlated, or belong together, may, after prolonged admixture with another race, pass down separately, and form what is known as disharmonic combinations. Such disharmonic combinations are, for example, a tall brunet, or a short blond; blue eyes associated with brunet hair, or brown eyes with blond hair. In modern science the meaning of the word "character" is now limited to physical instead of mental and spiritual traits as in popular usage.

The process of intermixture of unit characters has gone far in existing populations, and with the ease of modern methods of transportation this process is going much further in Europe, and in America. The immediate results of such mixture are not blends, or intermediate types, but rather mosaics of contrasted characters. Such blends, if any, as ultimately occur, are too remote to concern us here. The first result of the crossing of a pure brunet with a pure blond is to produce either pure blonds or pure brunets in certain known proportions, instead of offspring of an intermediate type; or else a third group which may be either blond or brunet, but which possesses latent characters of the contrasted type. Such latent or recessive characters often reappear in remote descendants.

In defining race in Europe it is necessary not only to consider pure groups or pure types, but also the distribution of unit characters belonging to each particular subspecies of man found there. The interbreeding of these populations has progressed to such an extent that in many cases such an analysis of physical characters is necessary to reconstruct the elements which have entered into their ethnic composition.

Sometimes we find a unit character appearing here and there as the sole remnant of a once numerous race, for example, the occasional appearance in European populations of a skull of the Neanderthal type, a race widely spread over Europe 40,000 years ago, or of the Cro-Magnon type, the predominant race 16,000 years ago. Before the fossil remains of the Neanderthal and Cro-Magnon races were studied and understood such reversional specimens were considered pathological, instead of being recognized as the reappearance of an ancient and submerged type.

Unit characters are to all intents and purposes immutable, and they do not change during the lifetime of a language, or an empire. The skull shape of the Egyptian fellaheen, in the unchanging environment of the Nile Valley, is absolutely identical in measurements, proportions and capacity with skulls found in the predynastic tombs dating back more than six thousand years.

There exists to-day a widespread and fatuous belief in the power of environment, as well as of education and opportunity to alter heredity, which arises from the dogma of the brotherhood of man, derived in turn from the loose thinkers of the French Revolution and their American mimics. Such beliefs have done much damage in the past, and if allowed to go uncontradicted, may do much more serious damage in the future. Thus the view that the negro slave was an unfortunate cousin of the white man, deeply tanned by the tropic sun, and denied the blessings of Christianity and civilization, played no small part with the sentimentalists of the Civil War period, and it has taken us fifty years to learn that speaking English, wearing good clothes, and going to school and to church, does not transform a negro into a white man. Nor was a Syrian or Egyptian freedman transformed into a Roman by wearing a toga, and applauding his favorite gladiator in the amphitheatre. We shall have a similar experience with the Polish Jew, whose dwarf stature, peculiar mentality, and ruthless concentration on self-interest are being engrafted upon the stock of the nation.

Recent attempts have been made in the interest of inferior races among our immigrants to show that the shape of the skull does change, not merely in a century, but in a single generation.

In 1910, the report of the anthropological expert of the Congressional

Immigration Commission, gravely declared that a round skull Jew on his way across the Atlantic might and did have a round skull child, but that a few years later, in response to the subtle elixir of American institutions, as exemplified in an East Side tenement, might and did have a child whose skull was appreciably longer; and that a long skull south Italian, breeding freely, would have precisely the same experience in the reverse direction. In other words, the Melting Pot was acting instantly under the influence of a changed environment.

What the Melting Pot actually does in practice, can be seen in Mexico, where the absorption of the blood of the original Spanish conquerors by the native Indian population has produced the racial mixture which we call Mexican, and which is now engaged in demonstrating its incapacity for self-government. The world has seen many such mixtures of races, and the character of a mongrel race is only just beginning to be understood at its true value.

It must be borne in mind that the specializations which characterize the higher races are of relatively recent development, are highly unstable and when mixed with generalized or primitive characters, tend to disappear. Whether we like to admit it or not, the result of the mixture of two races, in the long run, gives us a race reverting to the more ancient, generalized and lower type. The cross between a white man and an Indian is an Indian; the cross between a white man and a negro is a negro; the cross between a white man and a Hindu is a Hindu; and the cross between any of the three European races and a Jew is a Jew.

In the crossing of the blond and brunet elements of a population, the more deeply rooted and ancient dark traits are prepotent or dominant. This is matter of everyday observation, and the working of this law of nature is not influenced or affected by democratic institutions or by religious beliefs.

As measured in terms of centuries, unit characters are immutable, and the only benefit to be derived from a changed environment and better food conditions, is the opportunity afforded a race which has lived under adverse conditions, to achieve its maximum development, but the limits of that development are fixed for it by heredity and not by environment.

In dealing with European populations the best method of determining race has been found to lie in a comparison of proportions of the skull, the so-called cephalic index. 'This is the ratio of maximum *length* to maximum *width* taken at the widest part of the skull above the ears. Skulls with an index of 75 or less, that is, when the width is three-fourths or less than the length, are considered dolichocephalic, or long skulls. Skulls of an index of 80 or over are round skulls, or brachycephalic. Intermediate indices, between 75 and 80, are considered mesocephalic. These are cranial indices. To allow for the flesh on living specimens, about two per cent is to be added to the index, and the result is the cephalic index. In the following pages only long and round skulls are considered and the intermediate forms, or mesocephs, are assigned to the dolichocephalic group.

This cephalic index, though an extremely important if not the controlling unit character, is, nevertheless, but a single character and must be checked up with other somatological traits. Normally, a long skull is associated with a long face, and a round skull with a round face.

The use of this test, the cephalic index, enables us to divide the great bulk of the European populations into three distinct subspecies of man, one northern and one southern, both dolichocephalic or characterized by a long skull, and a central subspecies which is brachycephalic, or characterized by a round skull.

The first is the Nordic or Baltic subspecies. This race is long skulled, very tall, fair skinned, with blond or brown hair and light colored eyes. The Nordics inhabit the countries around the North and Baltic Seas, and include not only the great Scandinavian and Teutonic groups, but also other early peoples who first appear in southern Europe and in Asia as representatives of Aryan language and culture.

The second is the dark Mediterranean or Iberian subspecies, occupying the shores of the inland sea, and extending along the Atlantic coast until it reaches the Nordic species. It also spreads far east into southern Asia. It is long skulled like the Nordic race, but the absolute size of the skull is less. The eyes and hair are very dark or black, and the skin more or less swarthy. The stature is stunted in comparison to that of the Nordic race and the musculature and bony framework weak.

The third is the Alpine subspecies occupying all central and eastern Europe, and extending through Asia Minor to the Hindu Kush and the Pamirs. The Armenoids constitute an Alpine subdivision and represent the ancestral type of this race which remained in the mountains and high plateaux of Anatolia and western Asia. The Alpines are round skulled, of medium height and sturdy build, both as to skeleton and muscles. The coloration of both hair and eyes was originally very dark and still tends strongly in that direction, but many light colored eyes, especially gray, are now found in the Alpine populations of western Europe.

While the inhabitants of Europe betray as a whole their mixed origin, nevertheless the three main subspecies are each found in large numbers and in great purity, as well as sparse remnants of still more ancient races represented by small groups or by individuals, and even by unit characters.

These three main groups have bodily characters which constitute them distinct subspecies of *Homo sapiens*. Each has several varieties, but for the sake of clearness the word race and not the word species or subspecies will hereafter be used nearly, but not quite, exclusively. In zoology the term species implies the existence of a certain definite amount of divergence from the most closely related type, but race does not require a similar amount of difference. In man, where all groups are more or less fertile when crossed, so many intermediate or mixed types occur that the word species has too limited a meaning for wide use. Related species when grouped together constitute subgenera and genera.

The old idea that fertility or infertility of races of animals was the measure of species, is now abandoned. One of the greatest difficulties in classifying man is his perverse predisposition to mismate. This is a matter of daily observation, especially among the women of the better classes, probably because of their wider range of choice.

The cephalic index is of less value in the classification of Asiatic populations, but the distribution of round and long skulls is similar to that in Europe. The vast central plateau of that continent is inhabited by round skulls. In fact, Thibet and the western Himalayas were probably the centre of radiation of all the round skulls of the world. In India and Persia south of this central area occurs a long skull race related to Mediterranean man in Europe.

Both skull types occur, much intermixed, among the American Indians, and the cephalic index is of little value in classifying the Amerinds. No satisfactory explanation of the variability of the skull shape of this species has as yet been found, but the total range of variation of physical characters from northern Canada to southern Patagonia is less than the range of such variation from Normandy to Provence in France.

In Africa the cephalic index is also of small classification value because all of the populations I are characterized by a long skull.

The distinction between a long skull and a x round skull in mankind probably goes back at least to early Paleolithic times, if not to a period still more remote. It is of such great antiquity that when new species or races appear in Europe at the close of the Paleolithic, between 10,000 and 7,000 years B. C., the skull characters among them are as clearly defined as they are to-day.

The fact that two distinct species of mankind both have long skulls, as have the north European and the Airican negro, is no necessary indication of relationship, and in that instance is merely a case of parallel specialization. The fact, however, that the Swede has a long skull and the Savoyard a round skull does prove them to be descendants of distinct subspecies.

The claims that the Nordic race is a mere variation of the Mediterranean race, and that the latter is, in turn, derived from the Ethiopian negro, rest upon a mistaken idea that a dolichocephaly in common must mean identity of origin, as well as upon a failure to take into consideration many somatological characters of almost equal value with the cephalic index. In this connection it is well to remark that this measurement, being merely a ratio, may yield identical figures for skulls differing in every other proportion and detail, as well as in absolute size and capacity.

Eye color is of very great importance in race determination, because all blue, gray, or green eyes in the world to-day came originally from the same source, namely, the Nordic race of northern Europe. This light colored eye has appeared nowhere else on earth, and is a specialization of this subspecies of man only, and is consequently one of extreme value in the classification of European races. Dark colored eyes are all but universal among wild mammals, and entirely so among the primates, man's nearest relatives. It is, therefore, an absolute certainty that all the original races of man had dark eyes.

One subspecies of man, and one alone, specialized in light colored eyes. This same subspecies also evolved light or blond hair, a character far less deeply rooted than eye color, as blond children tend to grow darker with advancing years, and populations largely of Nordic extraction, such as those of Lombardy, upon admixture with darker races, lose their blond hair more readily than their light colored eyes.

Blond hair also comes everywhere from the Nordic species, and from nowhere else. Whenever we find blondness among the darker races of the earth we may be sure some Nordic wanderer has passed that way. When individuals of perfect blond type occur, as sometimes in Greek islands, we may suspect a recent visit of sailors from a passing ship, but when only single characters remain spread thinly, but widely, over considerable areas, like the blondness of the Atlas Berbers or of the Albanian mountaineers, we must search in the dim past for the origin of these blurred traits of early invaders.

The range of blond hair color in pure Nordic peoples runs from flaxen and red to shades of chestnut and brown. The darker shades may indicate crossing in some cases, but absolutely black hair certainly does mean an ancestral cross with a dark race—in England with the Mediterranean race.

In Nordic populations the women are, in general, lighter haired than the men, a fact which points to a blond past and a darker future for those populations. Women in all human races, as the females among all mammals, tend to exhibit the older, more generalized and primitive traits of the race's past. The male in his individual development indicates the direction in which the race is tending under the influence of variation and selection.

It is interesting to note in connection with the more primitive physique of the female, that in the spiritual sphere also, women retain the ancient and intuitive knowledge that the great mass of mankind is not free and equal, but bond and unequal.

The color of the skin is a character of importance, but one that is exceedingly hard to measure as the range of variation in Europe between skins of extreme fairness and those that are exceedingly swarthy, is almost complete. In general the Nordic race in its purity has an absolutely fair skin, and is consequently the Homo albus, the white man par excellence.

Many members of the Nordic race otherwise apparently pure have skins, as well as hair, more or less dark, so that the determinative value of this character is uncertain. There can be no doubt that the quality of the skin and the extreme range of its variation in color from black, brown, red, yellow to ivory-white are excellent measures of the specific or subgeneric distinctions between the larger groups of mankind, but in dealing with European populations it is sometimes difficult to correlate shades of fairness with other physical characters.

It often happens that an individual with all the Nordic characters in great purity, has a skin of an olive or dark tint, and it much more frequently happens that we find an individual with absolutely pure brunet traits in possession of a skin of almost ivory whiteness and of great clarity. This last combination is very frequent among the brunets of the British Isles. That these are, to some extent, disharmonic combinations we may be certain, but beyond that our knowledge does not lead. Owners, however, of a fair skin have always been, and still are, the objects of keen envy by those whose skins are black, yellow, or red.

Stature is another unit character of greater value than skin color, and perhaps than hair color, iand is one of much importance in European classification because on that continent we have the most extreme variations of human height.

Exceedingly adverse economic conditions may inhibit a race from attaining the full measure of its growth, and to this extent environment plays its part in determining stature, but fundamentally it is race, always race, that sets the limit. The tall Scot and the dwarfed Sardinian owe their respective sizes to race, and not to oatmeal or olive oil. It is probable that the fact that the stature of the Irish is, on the average, shorter than that of the Scotch, is due partly to economic conditions, and partly to the depressing effect of a considerable population of primitive short stock.

Mountaineers all over the world tend to be tall and vigorous, a fact probably due to the rigid elimination of defectives by the unfavorable environment. In this case altitude would operate like latitude, and produce the severe conditions which seem essential to human vigor. The short stature of the Lapps and the Esquimaux may have been originally attributable to the trying conditions of an Arctic habitat, but in any event it has long since become a racial character.

So far as the main species of Europe are concerned, stature is a very valuable measure of race.

To recapitulate as to this character, the Mediterranean race is everywhere marked by a relatively short stature, sometimes greatly depressed, as in south Italy and in Sardinia, and also by a comparatively light bony framework and feeble muscular development.

The Alpine race is taller than the Mediterranean although shorter than the Nordic, and is characterized by a stocky and sturdy build.

The Nordic race is nearly everywhere distinguished by great stature. Almost the tallest stature in the world is found among the pure Nordic populations of the Scottish and English borders, while the native British of Pre-Nordic brunet blood are, for the most part, relatively short; and no one can question the race value of stature who observes on the streets of London the contrast between the Piccadilly gentleman of Nordic race and the cockney costermonger of the old Neolithic type.

In many cases where these three European races have become mixed, stature seems to be one of the first Nordic characters to vanish, but wherever in Europe we find great stature in a population otherwise lacking in Nordic characters, we may be certain of Nordic crossing, as in the case of a large proportion of the inhabitants of Burgundy, of Switzerland, of the Tyrol, and of the Dalmatian Alps south to Albania.

These four unit characters, skull shape, eye color, hair color, and stature, are sufficient to enable us to differentiate clearly between the three main races of Europe, but if we wish to discuss the minor variations and mixtures, we would have to go much further and take up other proportions of the skull than the cephalic index, as well as the shape and position of the eyes, and the proportions and shape of the jaws and chin.

The nose also is an exceedingly important character. The original human nose was, of course, broad and bridgeless. This trait is shown clearly) in new-born infants who recapitulate in their development the various stages of the evolution of the human genus. A bridgeless nose with wide flaring nostrils is a very primitive character, and is still retained by some of the larger divisions of mankind throughout the world. It appears occasionally in white populations of European origin, but is everywhere a very ancient, generalized, and low character.

The high bridge and long, narrow nose, the so-called Roman, Norman, or aquiline nose, is characteristic of the most highly specialized races of mankind. While an apparently unimportant character, the nose is one of the very best clews to racial origin, and in the details of its form, and especially in the lateral shape of the nostrils, is a race determinant of the greatest value.

The lips, whether thin or fleshy or whether clean-cut or everted, are race characters. Thick, protruding, everted lips are very ancient traits and are characteristic of primitive races. A high in-step also has long been esteemed an indication of patrician type, while the flat foot is often the test of lowly origin.

The absence or abundance of hair and beard and the relative absence or abundance of body hair are characters of no little value in classification. Abundant body hair is, to a large extent, peculiar to populations of the very highest as well as the very lowest species, being characteristic of the north European as well as of the Australian savages. It merely means the retention in both these groups of a very early and primitive trait which has been lost by the Negroes, Mongols, and the Amerinds.

The Nordic and Alpine races are far better equipped with head and body hair than the Mediterranean, which is throughout its range a glabrous or relatively naked race.

The so-called red haired branch of the Nordic race has special characters in addition to red hair, such as a greenish cast of eye, a skin of peculiar texture tending either to great clarity or to freckles, and certain peculiar temperamental traits. This was probably a variety closely related to the blonds, and it first appears in history in association with them.

In the structure of the head hair of all races of mankind we find a regular progression from extreme kinkiness to lanky straightness, and this straightness or curliness depends on the shape of the cross section of the hair itself. This cross section has three distinct forms, corresponding with the most extreme divergences among human species.

While the three main European races are the subject of this book, and while it is not the intention of the author to deal with the other human types, it is necessary at this point to state that these three European subspecies, are subdivisions of one of the primary groups or subgenera of the genus Homo which, taken together, we must call the Caucasian for lack of a better name.

The great mass of the rest of mankind can be roughly divided into the Negroes and Negroids, and the Mongols and Mongoloids.

The former apparently originated in south Asia and entered Africa from the northeasterly corner of that continent. Africa south of the Sahara is now the chief home of this race, though remnants of Negroid aborigines are found throughout south Asia from India to the Philippines, while the very distinct black Melanesians and the Australoids lie farther to the east and south.

A third subgenus of mankind includes the round skulled Mongols and their derivatives, the Amerinds, or American Indians. This group is essentially Asiatic, and occupies the centre and the eastern half of that continent. A description of these Negroid and Mongoloid subgenera and their derivatives, as well as of certain aberrant species of man, lies outside of the scope of this work.

In the consideration of this measurement, the cross section of the hair in connection with these main subgenera, we find that a permanent relation exists, and that each of the three primary divisions of mankind is, in the shape of the cross section of its hair, differentiated from the others.

The cross section of the hair of the Negro and Negroid races is a flat ellipse with the result that all the members of this subgenus have kinky hair.

The cross section of the hair of the Mongols and their derivatives, the Amerinds, is a complete circle, and the hair of this subgenus is perfectly straight and lank.

The cross section of the hair of the so-called Caucasians, including the Mediterranean, Alpine, and Nordic subspecies, is an oval ellipse, and consequently is intermediate between the cross sections of the Negroids and Mongoloids. Hair of this structure is wavy or curly, never either kinky or absolutely straight, and is characteristic of all the European populations, almost without exception.

We have confined our discussion to the most important unit characters, but there are many other valuable aids to classification to be found in the proportions of the body and the relative length of the limbs. For an example, it is a matter of common knowledge that there occur among white women two distinct types in this latter respect, the one long legged and short bodied, the other long bodied and short legged. All such facts have a race value as yet not understood.

Without going into further physical details, it is probable that all relative proportions in the body, the features, the skeleton, and the skull which are fixed and constant and lie outside of the range of individual variation represent dim inheritances from the past. Every human being unites in himself the blood of thousands of ancestors, stretching back through thousands of years, superimposed upon a pre-human inheritance of still greater antiquity, and the face and body of every living man offer an intricate mass of hieroglyphs that science will some day learn to read and interpret.

We shall use the foregoing main unit characters as the basis of our definition of race, and shall later call attention to such temperamental and spiritual traits as seem to be associated with distinct physical types.

We shall only discuss European populations and shall not deal with those quarters of the globe where the races of man are such that other physical characters must be called upon to provide clear definitions.

A fascinating subject would open up if we were to dwell upon the effect of racial combinations and disharmonies, as, for instance, where the mixed Nordic and Alpine populations of Lombardy retain the skull shape, hair color, and stature of the Alpine race, with the light eye color of the Nordic race, or where the mountain populations along the east coast of the Adriatic from the Tyrol to Albania have the stature of the Nordic race and an Alpine skull and coloration.

# PART III

# RACE AND HABITAT

The laws which govern the distribution of the various races of man and their evolution through selection are substantially the same as those controlling the evolution and distribution of the larger mammals.

Man, however, with his superior mentality, has freed himself from many of the elements which impose restraint upon the expansion of animals. In his case selection through disease and social and economic competition has replaced selection through adjustment to the limitations of food supply.

Man is the most cosmopolitan of animals, and in one form or another thrives in the tropics and in the arctics, at sea level and on high plateaux, in the desert and in the reeking forests of the equator. Nevertheless, the various races of Europe with which we deal in this book have, each of them, a certain natural habitat in which each achieves its highest development.

### The Nordic Habitat

The Nordics appear in their present centre of distribution, the basin of the Baltic, -at the close of the Paleolithic, as soon as the retreating glaciers left habitable land. This race was probably at that time in possession of its fundamental characters, and its extension in the Teutonic group from the plains of Russia to Scandinavia was not in the nature of a radical change of environment. The race in consequence is now and always has been, probably always will be, adjusted to certain environmental conditions, chief of which is protection from a tropical sun. The actinic rays of the sun I at the same latitude are uniform in strength the world over, and continuous sunlight affects adversely the delicate nervous organization of the Nordics. The fogs and long winter nights of the North serve as a protection from too much sun, and from its too direct rays.

Scarcely less important is the presence of a large amount of moisture, but above all a constant variety of temperature is needed. Sharp contrast between night and day temperature, and between summer and winter are necessary to maintain the vigor of the blond race at a high pitch. Uniform weather, if long continued, lessens its energy. Too great extremes, as in midwinter or midsummer in New England, are injurious. Limited but constant alternations of heat and cold, of moisture and dryness, of sun and clouds, of calm and cyclonic storms, offer the ideal surroundings for the Nordic race.

Men of the Nordic race may not enjoy the fogs and snows of the North, the endless changes of weather, and the violent fluctuations of the thermometer, and they may seek the sunny southern isles, but under the former conditions they flourish, do their work, and raise their families. In the south they grow listless and cease to breed.

In the lower classes the increasing proportion of poor whites and "crackers" are symptoms of lack of climatic adjustment. The whites in Georgia, the Bahamas, and above all the Barbadoes are excellent examples of the deleterious effects of residence outside the natural habitat of the Nordic race.

The poor whites of the Cumberland Mountains in Kentucky and Tennessee present a more difficult problem, because here the altitude, even though small, should modify the effects of latitude, and the climate of these mountains cannot be particularly unfavorable to men of Nordic breed. There are probably other hereditary forces at work here as yet little understood.

No doubt bad food and economic conditions, prolonged inbreeding, and the loss through emigration of the best elements have played a large part in the degeneration of these poor whites. They represent to a large extent the offspring of bond servants brought over by the rich planters in early Colonial times. Their names indicate that, many of them are the descendants of the old borderers along the Scotch and English frontier, and the persistence with which family feuds are maintained certainly points to such an origin. The physical type is typically Nordic, for the most part pure Saxon or Anglian, and the whole mountain population show somewhat aberrant but very pronounced physical, moral, and mental characteristics which would repay scientific investigation. The problem is too complex to be disposed of by reference to the hookworm, illiteracy, or competition with negroes.

This type played a very large part in the settlement of the Middle West, by way of Kentucky, Tennessee, and Missouri. Thence they passed both up the Missouri River and down the Santa Fe trail, and contributed rather more than their share of the train robbers, horse thieves, and bad men of the West.

Scotland and the Bahamas are inhabited by men of precisely the same race, but the vigor of the English in the Bahamas is gone, and the beauty of their women has faded. The fact that they were not in competition with an autochthonous race better adjusted to climatic conditions has enabled them to survive, but the type could not have persisted, even during the last two hundred years, if they had been compelled to compete on terms of equality with a native and acclimated population.

Another element entering into racial degeneration on many other islands, and for that matter in many New England villages, is the loss through emigration of the more vigorous and energetic individuals, leaving behind the less efficient to continue the race at home.

In subtropical countries, when the energy of the Nordics is at a low ebb, it would appear that the racial inheritance of physical strength and mental vigor were suppressed and recessive rather than destroyed. Many individuals who were born in unfavorable climatic surroundings, but who move back to the original habitat of their race in the north, recover their full quota of energy and vigor. New York and other Northern cities have many Southerners who are fully as efficient as pure Northerners.

This blond race can exist outside of its native environment as land owning aristocrats who are not required to do manual labor in the fields under a blazing sun. As such an aristocracy it continues to exist under Italian skies, but as a field laborer the man of Nordic blood could not compete with his Alpine or Mediterranean rival. It is not to be supposed that the Teutonic armies which for a thousand years after the fall of Rome poured down from the Alps like the glaciers to melt in the southern sun, were composed solely of knights and gentlemen who became the landed nobility of Italy. The man in the ranks also took up his land and work in Italy, but he had to compete directly with the native under climatic conditions which were unfavorable to his race. In this competition the blue eyed Nordic giant died, and the native survived. His officer, however, lived in the castle and directed the labor of his bondsmen without other preoccupation than the chase and war, and he long maintained his vigor.

The same thing happened in our South before the Civil War. There the white men did not work in the fields or in the factory. The heavy work under the blazing sun was performed by negro slaves, and the planter was spared exposure to an unfavorable environment. Under these conditions he was able to retain much of his vigor. When slavery was abolished, and the white man had to plough his own fields or work in the factory, deterioration began.

The change in the type of men who are now sent by the Southern States to represent them in the Federal Government from their predecessors in ante-bellum times is partly due to these causes, but in a greater degree it is to be attributed to the fact that a very large portion of the best racial strains in the South were killed off during the Civil War. In addition the war shattered the aristocratic traditions which formerly secured the selection of the best men as rulers. The new democratic ideals with universal suffrage in free operation among the whites result in the choice of representatives who lack the distinction and ability of the leaders of the Old South.

A race may be thoroughly adjusted to a certain country at one stage of its development and be at a disadvantage when an economic change occurs, such as was experienced in England a century ago when the nation changed from an agricultural to a manufacturing community. The type of man that flourishes in the fields is not the type of man that thrives in the factory, just as the type of man required for the crew of a sailing ship is not the type useful as stokers on a modern steamer.

## The Habitat of the Alpines and Mediterraneans

The environment of the Alpine race seems to have always been the mountainous country of central and eastern Europe, as well as western Asia. This type has never flourished in the deserts of Arabia or the Sahara, nor has it succeeded in maintaining its colonies in the north of Europe within the domain of the Nordic long heads. It is, however, a sturdy and persistent stock, and, while much of it may not be over-refined or cultured, undoubtedly possesses great potentialities for future development.

The Alpines in the west of Europe, especially in Switzerland and the districts immediately surrounding, have been so thoroughly Nordicized, and so saturated with the culture of the adjoining nations, that they stand in sharp contrast to backward Alpines of Slavic speech, in the Balkans and east of Europe.

The Mediterranean race, on the other hand, is clearly a southern type with eastern affinities. It is a type that did not flourish in the north of Europe under old agricultural conditions, nor is it suitable to the farming districts and frontiers of America and Canada. It is adjusted to subtropical and tropical countries better than any other European type, and will flourish in our Southern States and around the coasts of the Spanish Main. In France it is well known that members of the Mediterranean race are better adapted for colonization in Algeria than are French Alpines or Nordics. This subspecies of man is notoriously intolerant of extreme cold, owing to its sensibility to diseases of the lungs, and it shrinks from the blasts of the northern winter in which the Nordics revel.

The brunet Mediterranean element in the native American seems to be increasing at the expense of the blond Nordic element generally throughout the Southern States, and probably also in the large cities. This type of man, however, is scarce on our frontiers. In the Northwest, and in Alaska in the days of the gold rush, it was in the mining camps a matter of comment if a man turned up with dark eyes, so universal were blue and gray eyes among the American pioneers.

# IV
# THE COMPETITION OF RACES

Where two races occupy a country side by side, it is not correct to speak of one type as changing into the other. Even if present in equal numbers one of the two contrasted types will have some small advantage or capacity which the other lacks toward a perfect adjustment to surroundings. Those possessing these favorable variations will flourish at the expense of their rivals, and their offspring will not only be more numerous, but will also tend to inherit such variations. In this way one type gradually breeds the other out. In this sense, and in this sense only, do races change.

Man continuously undergoes selection through social environment. Among native Americans of the Colonial period a large family was an asset, and social pressure and economic advantage both counselled early marriage and numerous children. Two hundred years of continuous political expansion and material prosperity changed these conditions and children, instead of being an asset to till the fields and guard the cattle, became an expensive liability. They now require support, education, and endowment from their parents, and a large family is regarded by some as a serious handicap in the social struggle.

These conditions do not obtain at first among immigrants, and large families among the newly arrived population are still the rule, precisely as they were in Colonial America, and are to-day in French Canada, where backwoods conditions still prevail.

The result is that one class or type in a population expands more rapidly than another, and ultimately replaces it. This process of replacement of one type by another does not mean that the race changes, or is transformed into another. It is a replacement pure and simple and not a transformation.

The lowering of the birth rate among the most valuable classes, while the birth rate of the lower classes remains unaffected, is a frequent phenomenon of prosperity. Such a change becomes extremely injurious to the race if unchecked; unless nature is allowed to maintain by her own cruel devices the relative numbers of the different classes in their due proportions. To attack race suicide by encouraging indiscriminate breeding is not only futile, but is dangerous if it leads to an increase in the undesirable elements. What is needed in the community most of all, is an increase in the desirable classes, which are of superior type physically, intellectually, and morally, and not merely an increase in the absolute numbers of the population.

The value and efficiency of a population are not numbered by what the newspapers call souls, but by the proportion of men of physical and intellectual vigor. The small Colonial population of America was, man for man, far superior to the average of the present inhabitants, although the latter are twenty-five times more numerous. The ideal in eugenics toward which statesmanship should be directed, is, of course, improvement in quality rather than quantity. This, however, is at present a counsel of perfection, and we must face conditions as they are.

The small birth rate in the upper classes is, to some extent, offset by the care received by such children as are born, and the better chance they have to become adult and breed in their turn. The large birth rate of the lower classes is, under normal conditions, offset by a heavy infant mortality, which eliminates the weaker children.

Where altruism, philanthropy, or sentimentalism intervene with the noblest purpose, and forbid nature to penalize the unfortunate victims of reckless breeding, the multiplication of inferior types is encouraged and fostered. Efforts to indiscriminately preserve babies among the lower classes often result in serious injury to the race.

Mistaken regard for what are believed to be divine laws and a sentimental belief in the sanctity of human life, tend to prevent both the elimination of defective infants and the sterilization of such adults as are themselves of no value to the community. The laws of nature require the obliteration of the unfit, and human life is valuable only when it is of use to the community or race.

It is highly unjust that a minute minority should be called upon to supply brains for the unthinking mass of the community, but it is even worse to burden the responsible and larger, but still overworked, elements in the community with an ever increasing number of moral perverts, mental defectives, and hereditary cripples.

The church assumes a serious responsibility toward the future of the race whenever it steps in and preserves a defective strain. The marriage of deaf mutes was hailed a generation ago as a triumph of humanity. Now it is recognized as an absolute crime against the race. A great injury is done to the community by the perpetuation of worthless types. These strains are apt to be meek and lowly, and as such make a strong appeal to the sympathies of the successful. Before eugenics were understood much could be said from a Christian and humane view-point in favor of indiscriminate charity for the benefit of the individual. The societies for charity, altruism, or extension of rights, should have, however, in these days, in their management some small modicum of brains, otherwise they may continue to do, as they have sometimes done in the past, more injury to the race than black death or smallpox.

As long as such charitable organizations confine themselves to the relief of suffering individuals, no matter how criminal or diseased they may be, no harm is done except to our own generation, and if modern society recognizes a duty to the humblest malefactors or imbeciles, that duty can be harmlessly performed in full, provided they be deprived of the capacity to procreate their defective strain.

Those who read these pages will feel that there is little hope for humanity, but the remedy has been found, and can be quickly and mercifully applied. A rigid system of selection through the elimination of those who are weak or unfit—in other words, social failures—would solve the whole question in one hundred years, as well as enable us to get rid of the undesirables who crowd our jails, hospitals, and insane asylums. The individual himself can be nourished, educated, and protected by the community during his lifetime, but the state through sterilization must see to it that his line stops with him, or else future generations will be cursed with an ever increasing load of victims of misguided sentimentalism. This is a practical, merciful, and inevitable solution of the whole problem, and can be applied to an ever widening circle of social discards, beginning always with the criminal, the diseased, and the insane, and extending gradually to types which may be called weaklings rather than defectives, and perhaps ultimately to worthless race types.

Efforts to increase the birth rate of the genius producing classes of the community, while most desirable, encounter great difficulties. In such efforts we encounter social conditions over which we have as yet no control. It was tried two thousand years ago by Augustus, and his efforts to avert race suicide and the extinction of the old Roman breed were singularly prophetic of what some far seeing men are attempting in order to preserve the race of native Americans of Colonial descent.

Man has the choice of two methods of race improvement. He can breed from the best, or he can eliminate the worst by segregation or sterilization. The first method was adopted by the Spartans, who had for their national ideals, military efficiency and the virtues of self control, and along these lines the results were completely successful. Under modem social conditions it would be extremely difficult in the first instance to determine which were the most desirable types, except in the most general way, and even if a satisfactory selection were finally made, it would be, in a democracy, a virtual impossibility to limit by law the right to breed to a privileged and chosen few.

Experiments in limiting breeding to the undesirable classes were unconsciously made in mediæval Europe under the guidance of the church. After the fall of Rome, social conditions were such that all those who loved a studious and quiet life, were compelled to seek refuge from the violence of the times in monastic institutions, and upon such the church imposed the obligation of celibacy, and thus deprived the world of offspring from these desirable classes.

In the Middle Ages, through persecution resulting in actual death, life imprisonment, and banishment, the free thinking, progressive, and intellectual elements were persistently eliminated over large areas, leaving the perpetuation of the race to be carried on by the brutal, the servile, and the stupid. It is now impossible to say to what extent the Roman Church by these methods has impaired the brain capacity of Europe, but in Spain alone, for a period of over three centuries, from the year 1471 to 1781, the Inquisition condemned to the stake or imprisonment an average of 1,000 persons annually. During these three centuries no less than 32,000 were burned alive, and 291,000 were condemned to various terms of imprisonment and other penalties, and 7,000 persons were burned in effigy, representing men who had died in prison or had fled the country.

No better method of eliminating the genius producing strains of a nation could be devised, and if such were its purpose the result was eminently satisfactory, as is demonstrated by the superstitious and unintelligent Spaniard of to-day. A similar elimination of brains and ability took place in northern Italy and in France, and in the Low Countries, where hundreds of thousands of Huguenots were murdered or driven into exile.

Under existing conditions the most practical and hopeful method of race improvement is through the elimination of the least desirable elements in the nation by depriving them of the power to contribute to future generations. It is well known to stock breeders that the color of a herd of cattle can be modified by continuous elimination of worthless shades, and of course this is true of other characters. Black sheep, for instance, have been practically destroyed by cutting out generation after generation all animals that show this color phase, until in carefully maintained flocks a black individual only appears as a rare sport.

In mankind it would not be a matter of great difficulty to secure a general consensus of public opinion as to the least desirable, let us say, ten per cent of the community. When this unemployed and unemployable-human residuum has been eliminated, together with the great mass of crime, poverty, alcoholism, and feeblemindedness associated therewith, it would be easy to consider the advisability of further restricting the perpetuation of the then remaining least valuable types. By this method mankind might ultimately become sufficiently intelligent to deliberately choose the most vital and intellectual strains to carry on the race.

In addition to selection by climatic environment, man is now, and has been for ages, undergoing selection through disease. He has been decimated throughout the centuries by pestilences such as the black death and bubonic plague. In our fathers' days yellow fever and smallpox cursed humanity. These plagues are now under control, but similar diseases, now regarded as mere nuisances to childhood, such as measles, mumps, and scarlatina, are terrible scourges to native populations without previous experience with them. Add to these smallpox and other white men's diseases, and one has the great empire builders of yesterday. It was not the swords in the hands of Columbus and his followers that decimated the American Indians, it was the germs that his men and their successors brought over, implanting the white man's maladies in the red man's world. Long before the arrival of the Puritans in New England, smallpox had flickered up and down the coast until the natives were but a broken remnant of their former numbers.

At the present time the Nordic race is undergoing selection through alcoholism, a peculiarly Nordic vice, and through consumption, and both these dread scourges unfortunately attack those members of the race that are otherwise most desirable, differing in this respect from filth diseases like typhus, typhoid, or smallpox. One has only to look among the more desirable classes for the victims of rum and tubercule to realize that death or mental and physical impairment through these two causes have cost the race many of its most brilliant and attractive members.

# V
# RACE, LANGUAGE, AND NATIONALITY

Nationality is an artificial political grouping of population, usually centering around a single language as an expression of traditions and aspirations. Nationality can, however, exist independently of language, but states thus formed, such as Belgium or Austria, are far less stable than those where a uniform language is prevalent, as, for example, France or England.

States without a single national language are constantly exposed to disintegration, especially 96 where a substantial minority of the inhabitants speak a tongue which is predominant in an adjoining state with, as a consequence, a tendency to gravitate toward such state.

The history of the last century in Europe has been the record of a long series of struggles to unite in one political unit all those speaking the same, or closely allied, dialects. With the exception of internal and social revolutions, every European war since the Napoleonic period has been caused by the effort to bring about the unification either of Italy or of Germany, or by the desperate attempts of the Balkan States to struggle out of Turkish chaos into modern European nations on a basis of community of language. The unification of both Italy and Germany is as yet incomplete, according to the views held by their more advanced patriots, and the solution of the Balkan question is still in the future.

Men are keenly aware of their nationality and are very sensitive about their language, but only in a few cases, notably in Sweden and Germany, does any large section of the population possess anything analogous to true race consciousness, although the term "race" is everywhere misused designate linguistic or political groups.

It sometimes happens that a section of the population of a large nation gathers around language, reinforced by religion, as an expression of individuality. The struggle between the French-speaking Alpine Walloons and the Nordic Flemings of Low Dutch tongue in Belgium is an example of two competing languages in an artificial nation which was formed originally around religion. On the other hand, the Irish National movement centers chiefly around religion, reinforced by myths of ancient grandeur. The French Canadians and the Poles use both religion and language to hold together what they consider a political unit. None of these so-called nationalities are founded on race.

During the past century alongside of the tendency to form imperial or large national groups, such as the Pan-Germanic, Pan-Slavic, Pan-Rumanian or Italia Irredenta movements, there has appeared a counter movement on the part of small disintegrating "nationalities" to reassert themselves, such as the Bohemian, Bulgar, Serb, Irish, and Egyptian national revivals. The upheaval is usually caused, as in the cases of the Irish and the Serbians, by delusions of former greatness now become national obsessions, but sometimes it means the resistance of a small group of higher culture to absorption by a lower civilization.

Examples of a high type threatened by a lower culture are afforded by the Finlanders, who are trying to escape the dire fate of their neighbors across the Gulf of Finland—the Russification of the Germans and Swedes of the Baltic Provinces—and by the struggle of the Danes of Schleswig to escape Germanization. The Armenians, too, have resisted stoutly the pressure of Islam to force them away from their ancient Christian faith. This people really represents the last outpost of Europe toward the Mohammedan East and constitutes the best remaining medium through which Western ideals and culture can be introduced into Asia.

In these as in other cases, the process of absorption from the viewpoint of the world at large is good or evil exactly in proportion to the relative value of the culture and race of the two groups.

The world would be no richer in civilization with an independent Bohemia or an enlarged Rumania, but, on the contrary, an independent Hungarian nation or an enlarged Greece would add greatly to the forces that make for good government and progress. An independent Ireland worked out on a Tammany model is not a pleasing prospect. A free Poland, apart from its value as a buffer state, would be actually a step backward. Poland was once great, but the elements that made it so are dead and gone, and to-day Poland is a geographical expression and nothing more.

The prevailing lack of true race consciousness is probably due to the fact that every important nation in Europe, as at present organized, with the sole exception of the Iberian and Scandinavian states, possesses in large proportions representatives of at least two of the fundamental European subspecies of man and of all manner of crosses between them. In France to-day, as in Caesar's Gaul, the three races divide the nation in almost equal proportions.

In the future, however, with an increased knowledge of the correct definition of true human species and types, and with a recognition of the immutability of fundamental racial characters, and of the results of mixed breeding, far more value will be attached to racial in contrast to national or linguistic affinities. In marital relations the consciousness of race will also play a much larger part than at present, although in the social sphere we shall have to contend with a certain strange attraction for contrasted types. When it becomes thoroughly understood that the children of mixed marriages between contrasted races belong to the lower type, the importance of transmitting in unimpaired purity the blood inheritance of ages will be appreciated at its full value, and to bring half-breeds into the world will be regarded as a social and racial crime of the first magnitude. The laws against miscegenation must be greatly extended if the higher races are to be maintained.

The language that a man speaks may be nothing more than evidence that at some time in the past his race has been in contact, either as conqueror or as conquered, with the original possessors of such language. One has only to consider the spread of the language of Rome over the vast extent of her empire, to realize how few of those who to-day speak Romance languages derive any portion of their blood from the pure Latin stock, and the error of talking about a "Latin race" becomes evident.

There is, however, such a thing as a large group of nations which have a mutual understanding and sympathy, based on the possession of a common or closely related group of languages and the culture of which it is the medium. This group may be called the "Latin nations," but never the "Latin race."

"Latin America" is a still greater misnomer as the great mass of the populations of South and Central America is not even European, and still less "Latin," being overwhelmingly of Amer-indian blood.

In the Teutonic group a large majority of those who speak Teutonic languages, as the English, Flemings, Dutch, North Germans, and Scandinavians, are descended from the Nordic race, and the dominant class in Europe is everywhere of that blood.

As to the so-called "Celtic race," the fantastic inapplicability of the term is at once apparent when we consider that those populations on the borders of the Atlantic Ocean, who to-day speak. Celtic dialects, are divided into three groups, each one showing in great purity the characters of one of the three entirely distinct human subspecies found in Europe. To class together the Breton peasant with his round Alpine skull; the little, long skull, brunet Welshman of the Mediterranean race, and the tall, blond, light eyed Scottish Highlander of pure Nordic race, in a single group labelled "Celtic," is obviously impossible. These peoples have neither physical, mental, nor cultural characteristics in common. If one be "Celtic" blood the other two clearly are not.

There was a people who were the original users of the Celtic language, and they formed the western vanguard of the Nordic race, which was spread all over central and western Europe, prior to the irruption of the Teutonic tribes. The descendants of these "Celts" must be sought today among those having the characters of the Nordic race and not elsewhere.

In England the little, dark Mediterranean Welshman talks about being Celtic quite unconscious that he is the residuum of Pre-Nordic races of immense antiquity. If the Celts are Mediterranean in race, then they are absent from central Europe, and we must regard as "Celts" all the Berbers and Egyptians, as well as many Persians and Hindus.

In France some enthusiasts regard the Breton of Alpine blood in the same light, and ignore his Asiatic origin. If these Alpine Bretons are "Celts" then there is not in the British Isles any substantial trace of their blood, as round skulls are practically absent there, and all the blond elements in England, Scotland, and Ireland must be attributed to the historic Teutonic invasions. Furthermore we must call all the continental Alpines "Celts," and must also include all Slavs, Armenians, and other brachycephs of western Asia within that designation, which would be obviously grotesque. The fact that the original Celts left behind their speech on the tongues of Mediterraneans in Wales, and of in Brittany, must not mislead us, as it indicates nothing more than that Celtic speech antedates the Teutons in England and the Romans in France. We must once and for all time discard the name "Celt" for any existing race whatever, and speak only of "Celtic" language and culture.

In Ireland the big, blond Nordic Danes, claim the honor of the name of "Celt," if honor it be, but the Irish are fully as Nordic as the English, the great mass of them being of Danish, Norse, and Anglo-Norman blood, in addition to earlier and PreTeutonic elements. We are all familiar with the blond and the brunet type of Irishman. These represent precisely the same racial elements as those which enter into the composition of the English, namely, the tall Nordic blond and the little Mediterranean brunet. The Irish are consequently not entitled to independent national existence on the ground of race, but if there is any ground for a political separation from England, it must rest, like that of Belgium, on religion, a basis for political combinations now happily obsolete in communities well advanced in culture.

In the case of the so-called "Slavic race," there is much more unity between racial type and language. It is true that in most Slavic-speaking countries the predominant race is clearly Alpine, except perhaps in Russia where there is a very large substratum of Nordic type—the so-called Finnic element, which may be considered as Proto-Nordic. The objection which is made to the identification of the Slavic race with the Alpine type rests chiefly on the fact that a very large portion of the Alpine race is German-speaking in Germany, Italian-speaking in Italy, and French-speaking in central France. Moreover, large portions of Rumania are of exactly the same racial complexion.

Many of the Greeks are also Alpines; in fact, are little more than Byzantinized Slavs. It was through the Byzantine Empire, that the Slavs first came in contact with the Mediterranean world, and through this Greek medium the Russians, the Serbians, the Rumanians, and the Bulgars received their Christianity.

Situated on the eastern marches of Europe the Slavs were submerged during long periods in the Middle Ages by Mongolian hordes, and were checked in development and warped in culture. Definite traces remain of the blood of the Mongols in both isolated and compact groups in south Russia, and scattered throughout the whole country as far west as the German boundary. The high tide of the Mongol invasion was during the thirteenth century. Three hundred years later the great Muscovite expansion began, first over the steppes to the Urals, and then across Siberian tundras and forests to the waters of the Pacific, taking up in its course much Mongolian blood, especially during the early stages of its advance.

The term "Caucasian race" has ceased to have any meaning except where it is used, in the United States, to contrast white populations with negroes or Indians, or, in the Old World, with Mongols. It is, however, a convenient term to include the three European subspecies when considered as divisions of one of the primary branches or subgenera of mankind. At best it is a cumbersome and archaic designation. The name "Caucasian" arose a century ago from a false assumption that the cradle of the blond Europeans was in the Caucasus, where there are now found no traces of any such race, except a small and decreasing minority of blond traits among the Ossetes, a tribe whose Aryan speech is related to that of the Armenians, and who, while mainly brachycephalic, still retain some blond and dolichocephalic elements which are apparently fading fast. The Ossetes have now about thirty per cent fair eyes and ten per cent fair hair. They are supposed to be, to some extent, a remnant of the Alans, a Teutonic tribe closely related to the Goths. Both Alans and Goths very early in our era occupied southern Russia, and were the latest known Nordics in the vicinity of the Caucasus Mountains. If these Ossetes are not partly of Alan origin they may possibly represent the last lingering trace of early Scythian dolichocephalic blondness.

The phrase "Indo-European race" is also of little use. If it has any meaning at all it must include all the three European races as well as members of the Mediterranean race in Persia and India. The use of this name also involves a false assumption of blood relationship between the main European populations and the Hindus, because of their possession in common of Aryan speech.

The name " Aryan race " must also be frankly discarded as a term of racial significance. It is to-day purely linguistic, although there was at one time, of course, an identity between the original Aryan mother tongue and the race that first spoke and developed it. In short there is not now, and there never was either a Caucasian or an Indo-European race, but there was once, thousands of years ago, an Aryan race now long since vanished into dim memories of the past. If used in a racial sense other than as above it should be limited to the Nordic invaders of Hindustan now long extinct. The great lapse of time since the disappearance of the ancient Aryan race as such, is measured by the extreme disintegration of the various groups of Aryan languages. These linguistic divergences are chiefly due to the imposition by conquest of Aryan speech upon several unrelated subspecies of man throughout western Asia and Europe.

# VI
# RACE AND LANGUAGE

When a country is invaded and conquered by a race speaking a foreign language, one of several things may happen, replacement of both population and language, as in the case of eastern England when conquered by the Saxons; or adoption of the language of the victors by the natives, as happened in Roman Gaul, where the invaders imposed their Latin tongue throughout the land, without substantially altering the race.

In England and Scotland later conquerors, Danes and Normans, failed to change the Saxon speech of the country, and in Gaul the German tongue of the Franks, Burgundians, and Northmen could not displace the language of Rome.

Autochthonous inhabitants frequently impose upon their invaders their own language and customs. In Normandy the conquering Norse pirates accepted the language, religion, and customs of the natives, and in a century they vanish from history as Scandinavian heathen and appear as the foremost representatives of the speech and religion of Rome.

In Hindustan the blond Nordic invaders forced their Aryan language on the aborigines, but their blood was quickly and utterly absorbed in the darker strains of the original owners of the land. A record of the desperate efforts of the conquering upper classes in India to preserve the purity of their blood persists until this very day in their carefully regulated system of castes. In our Southern States Jim Crow cars and social discriminations have exactly the same purpose and justification.

The Hindu to-day speaks a very ancient form of Aryan language, but there remains not one recognizable trace of the blood of the white conquerors who poured in through the passes of the Northwest. The boast of the modern Indian that he is of the same race as his English ruler, is entirely without basis in fact, and the little dark native

lives amid the monuments of a departed grandeur, professing the religion and speaking the tongue of his long forgotten Nordic conquerors, without the slightest claim to blood kinship. The dim and uncertain traces of Nordic blood in northern India only serve to emphasize the utter swamping of the white man in the burning South.

The power of racial resistance of a dense and thoroughly acclimated population to an incoming army, is very great. No ethnic conquest can be complete unless the natives are exterminated and the invaders bring their own women with them. If the conquerors are obliged to depend upon the women of the vanquished to carry on the race, the intrusive blood strain in a short time becomes diluted beyond recognition.

It sometimes happens that an infiltration of population takes place either in the guise of unwilling slaves, or of willing immigrants, filling up waste places and taking to the lowly tasks which the lords of the land despise, gradually occupying the country and literally breeding out their former masters.

The former catastrophe happened in the declining days of Rome, and the south Italians of to-day are very largely descendants of nondescript slaves of all races, chiefly from the southern and eastern coasts of the Mediterranean, who were imported by the Romans under the Empire to work their vast estates. The latter is occurring to-day in many parts of America, especially in New England.

The eastern half of Germany has a Slavic Alpine substratum which now represents the descendants of the Wends, who by the sixth century had filtered in as far west as the Elbe, occupying the lands left vacant by the Teutonic tribes which had migrated southward. These Wends in turn were Teutonized by a return wave of military conquest from the tenth century onward, and to-day their descendants are considered Germans in good standing. Having adopted the German as their sole tongue they are now in religious, political, and cultural sympathy with the pure Teutons;

in fact, they are quite unconscious of any racial distinction.

This historic fact underlies the ferocious controversy which has been raised over the ethnic origin of the Prussians, the issue being whether the populations in Brandenburg, Silesia, Posen, and other districts in eastern Germany, are Alpine Wends or true Nordic Germans. The truth is that the dominant half of the population is purely Teutonic and the lower half of the population are merely Teutonized Wends and Poles of Alpine affinities. Of course these territories must also retain some

of their early Teutonic population, and the blood of the Goth, Burgund, Vandal* and Lombard, who were at the commencement of our era located there, as well as the later Saxon element, must enter largely into the composition of the Prussian of to-day.

The most important communities in continental Europe of pure German type are to be found in old Saxony, the country around Hanover, and this element prevails generally in the northwestern part of the German Empire among the Low Dutch-speaking population, while the High German-speaking population is largely composed of Teutonized Alpines.'

All the states involved in the present world war have sent to the front their fighting Nordic element, and the loss of life now going on in Europe will fall much more heavily on the blond giant than on the little brunet.

As in all wars since Roman times, from a breeding point of view, the little dark man is the final winner. No one who saw one of our regiments march on its way to the Spanish War could fail to be impressed with the size and blondness of the men in the ranks as contrasted with the complacent citizen, who from his safe stand on the gutter curb gave his applause to the fighting man, and then stayed home to perpetuate his own brunet type.

This same Nordic element, everywhere the type of the sailor, the soldier, the adventurer, and the pioneer, was ever the type to migrate to new countries, until the ease of transportation and the desire to escape military service in the last forty years reversed the immigrant tide. In consequence of this change our immigrants now largely represent lowly refugees from "persecution" and other social discards.

In most cases the blood of pioneers has been lost to their race. They did not take their women with them. They either died childless or left half-breeds behind them. The virile blood of the Spanish conquistadores, who are now little more than a memory in Central and South America, died out from these causes.

This was also true in the early days of our Western frontiersmen, who individually were a far finer type than the settlers who followed them.

# VII
# THE EUROPEAN RACES IN COLONIES

For reasons already set forth there are few communities outside of Europe of pure European blood. The racial destiny of Mexico and of the islands and coasts of the Spanish Main is clear. The white man is being rapidly bred out by negroes on the islands and by Indians on the mainland. It is quite evident that the West Indies, the coast region of our Gulf States, and perhaps the black belt of the lower Mississippi Valley, must be abandoned to negroes. This transformation is already complete in Haiti, and is going rapidly forward in Cuba and Jamaica. Mexico and the northern part of South America must also be given over to native Indians with an ever thinning veneer of white culture of the "Latin" type.

In Venezuela the pure whites number about one per cent of the whole population, the balance being Indians and various crosses between Indians, negroes, and whites. In Jamaica the whites number not more than two per cent, while the remainder are negroes or mulattoes. In Mexico the proportion is larger, but the unmixed whites number not more than twenty per cent of the whole, the others being Indians pure or mixed. These latter are the "greasers" of the American frontiersman.

Whenever the incentive to imitate the dominant race is removed, the negro, or for that matter, the Indian, reverts shortly to his ancestral grade of culture. In other words, it is the individual .and not the race that is affected by religion, education, and example. Negroes have demonstrated throughout recorded time that they are a stationary species, and that they do not possess the potentiality of progress or initiative from within. Progress from self-impulse must not be confounded with mimicry or with progress imposed from without by social pressure, or by the slavers' lash.

Where two distinct species are located side by side history and biology teach that but one of two things can happen; either one race drives the other out, as the Americans exterminated the Indians, or as the negroes are now replacing the whites in various parts of the South; or else they amalgamate and form a population of race bastards in which the lower type ultimately preponderates. This is a disagreeable alternative with which to confront sentimentalists, but nature is only concerned with results and neither makes nor takes excuses. The chief failing of the day with some of our well meaning philanthropists is their absolute refusal to face inevitable facts, if such facts appear cruel.

In Argentine and south Brazil white blood of the various European races is pouring in so rapidly that a community preponderantly white, but of the Mediterranean type, may grow up, but such limited opportunities as the writer has had to observe Argentine types leads him to question the probability of such a result even there.

In Asia, with the sole exception of the Russian settlements in Siberia, there can be and will be no ethnic conquest, and all the white men in India, the East Indies, the Philippines, and China will leave not the slightest trace behind them in the blood of the native population. After several centuries of contact and settlement the pure Spanish in the Philippines are about half of one per cent. The Dutch in their East Indian islands are even less; while the resident whites in Hindustan amount to about one-tenth of one per cent. Such numbers are infinitesimal and of no force in a democracy, but in a monarchy, if kept free from contamination, they suffice for a ruling caste or a military aristocracy.

Australia and New Zealand, where the natives have been exterminated by the whites, are developing into communities of pure Nordic blood, and will for that reason play a large part in the future history of the Pacific. The bitter opposition of the Australians and Californians to the admission of Chinese coolies and Japanese farmers is due primarily to a blind but absolutely justified determination to keep those lands as white man's countries.

In Africa, south of the Sahara, the density of the native population will prevent the establishment of any purely white communities, except at the southern extremity of the continent and possibly on portions of the plateaux of eastern Africa. The stoppage of famines and wars and the abolition of the slave trade, while dictated by the noblest impulses of humanity, are suicidal to the white man. Upon the removal of these natural checks negroes multiply so rapidly that there will ( not be standing room on the continent for white I men, unless, perchance, the lethal sleeping sickness, far more fatal to blacks than to whites, should run its course unchecked.

In South Africa a community of Dutch and English extraction is developing. Here the only difference is one of language. English, being a world tongue, will inevitably prevail over the Dutch patois called "Taal." This Frisian dialect, as a matter of fact, is closer to old Saxon, or rather Kentish, than any living continental tongue, and the blood of the North Hollander is extremely close to that of the Anglo-Saxon of England. The English and the Dutch will merge in a common type just as they did two hundred years ago in the colony of New York. They must stand together if they are to maintain any part of Africa as a white man's country, because they are confronted with the menace of a large black Bantu population which will drive out the whites unless the problem is bravely faced.

The only possible solution is to establish large colonies for the negroes and to allow them outside of them only as laborers, and not as settlers. There must be ultimately a black South Africa and a white South Africa side by side, or else a pure black Africa from the Cape to the cataracts of the Nile.

In upper Canada, as in the United States up to the time of our Civil War, the white population was purely Nordic. The Dominion is, of course, handicapped by the presence of an indigestible mass of French-Canadians, largely from Brittany and of Alpine origin, although the habitant patois is an archaic Norman of the time of Louis XIV. These Frenchmen were granted freedom of language and religion by their conquerors, and are now using these privileges to form separatist groups in antagonism to the English population. The Quebec Frenchmen will succeed in seriously impeding the progress of Canada and will succeed even better in keeping themselves a poor and ignorant community of little more importance to the world at large than are the negroes in the South. The selfishness of the Quebec Frenchmen is measured by the fact that in the present war they will not fight for the British Empire, or for France, or even for clerical Belgium, and they are now endeavoring to make use of the military crisis to secure a further extension of their "nationalistic ideals."

Personally the writer believes that the finest and purest type of a Nordic community outside of Europe will develop in northwest Canada. Most of the other countries in which the Nordic race is now settling lie outside of the special environment in which alone it can flourish.

The negroes of the United States, while stationary, were not a serious drag on civilization until, in the last century, they were given the rights of citizenship and were incorporated in the body politic. These negroes brought with them no language or religion or customs of their own which persisted, but adopted all these elements of environment from the dominant race, taking the names of their masters just as to-day the German and Polish Jews are assuming American names. They came for the most part from the coasts of the Bight of Benin, but some of the later ones came from the southeast coast of Africa by way of Zanzibar. They were of various black tribes, but have been from the beginning saturated with white blood.

Looking at any group of negroes in America, it is easy to see that while they are all essentially negroes, whether coal black, brown, or yellow, the great majority of them have varying amounts of Nordic blood in them, which has modified their physical structure without transforming them in any way into white men. This miscegenation was, of course, a frightful disgrace to the dominant race, but its effect on the Nordics has been negligible, for the simple reason that it was confined to white men crossing with negro women, and not the reverse process, which would, of course, have resulted in the infusion of negro blood into the American stock.

The United States of America must be regarded racially as a European colony, and owing to current ignorance of the physical bases of race, one often hears the statement made that native Americans of Colonial ancestry are of mixed ethnic origin This is not true. At the time of the Revolutionary War the settlers in the thirteen Colonies were not only purely Nordic, but also purely Teutonic, a very large majority being Anglo-Saxon in the most limited meaning of that term. The New England settlers in particular came from those counties of England where the blood was almost purely Saxon, Anglian, and Dane.

New England, during Colonial times and long afterward, was far more Teutonic than old England; that is, it contained a smaller percentage of small, Pre-Nordic brunets. Any one familiar with the native New Englander knows the clean cut face, the high stature and the prevalence of gray and blue eyes and light brown hair, and recognizes that the brunet element is less noticeable there than in the South.

The Southern States were populated also by Englishmen of the purest Nordic type, but there is to-day, except among the mountains, an appreciably larger amount of brunet types than in the North. Virginia is in the same latitude as North Africa, and south of this line no blonds have ever been able to survive in full vigor, chiefly because the actinic rays of the sun are the same regardless of other climatic conditions. These rays beat heavily on the Nordic race and disturb their nervous system, wherever the white man ventures too far from the cold and foggy North.

The remaining Colonial elements, the Holland Dutch, the Palatine Germans, who came over in small numbers to New York and Pennsylvania, were also purely Teutonic, while the French Huguenots who escaped to America were drawn much more largely from the Nordic than from the Alpine or Mediterranean elements of France. The Scotch-Irish, who were numerous on the frontier of the middle Colonies were, of course, of pure Scotch and English blood, although they had resided in

Ireland two or three generations. They were quite free from admixture with the earlier Irish from whom they were cut off socially by bitter religious antagonism, and they are not to be considered as "Irish" in any sense.

There was no important immigration of other elements until the middle of the nineteenth century, when Irish Catholic and German immigrants appear for the first time upon the scene.

The Nordic blood was kept pure in the Colonies, because at that time among Protestant peoples there was a strong race feeling, as a result of which half-breeds between the white man and any native type were regarded as natives and not as white men.

There was plenty of mixture with the negroes as the light color of most negroes abundantly testifies, but these mulattoes, quadroons, or octoroons were then and are now universally regarded as negroes.

There was also abundant cross breeding along the frontiers between the white frontiersman and the Indian squaw, but the half-breed was everywhere regarded as a member of the inferior race.

In the Catholic colonies, however, of New France and New Spain, if the half-breed were a good Catholic he was regarded as a Frenchman or a Spaniard, as the case might be. This fact alone gives the clew to many of our colonial wars where the Indians, other than the Iroquois, were persuaded to join the French against the Americans by half-breeds who considered themselves Frenchmen. The Church of Rome has everywhere used its influence to break down racial distinctions. It disregards origins and only requires obedience to the mandates of the universal church. In that lies the secret of the opposition of Rome to all national movements. It is the imperial as contrasted with the nationalistic ideal, and in that respect the inheritance is direct from the Empire.

Race consciousness in the Colonies and in the j United States, down to and including the Mexican War, seems to have been very strongly developed! among native Americans, and it still remains in full vigor today in the South, where the presence of a large negro population forces this question upon the-' daily attention of the whites.

In New England, however, whether through the A decline of Calvinism or the growth of altruism, there appeared early in the last century a wave of sentimentalism, which at that time took up the cause of the negro, and in so doing apparently destroyed, to a large extent, pride and consciousness of race in the North. The agitation over slavery was inimical to the Nordic race, because it thrust aside all national opposition to the intrusion of hordes of immigrants of inferior racial value, and prevented the fixing of a definite American type, such as was clearly appearing in the middle of the century.

The Civil War was fought almost entirely by unalloyed native Americans. The German and Irish immigrants were at that time confined to a few States, and were chiefly mere day laborers and of no social importance. They played no part whatever in the development or policies of the nation, although in the war they contributed a certain number of soldiers to the Northern armies. These Irish and German elements were of Nordic race, and while they did not in the least strengthen the nation either morally or intellectually, they did not impair its physique.

There has been little or no Indian blood taken into the veins of the native American, except in States like Oklahoma and in some isolated families scattered here and there in the Northwest. This particular mixture will play no very important role in future combinations of race on this continent, except in the north of Canada.

The native American has always found, and finds now, in the black men, willing followers who ask only to obey and to further the ideals and wishes of the master race, without trying to inject into the body politic their own views, whether racial, religious, or social. Negroes are never socialists or labor unionists, and as long as the dominant imposes its will on the servient race, and as long as they remain in the same .relation to the whites as in the past, the negroes will be a valuable element in the community, but once raised to social equality their influence will be destructive to themselves and to the whites. If the purity of the races is to be maintained, they cannot continue to live side by side, and this is a problem from which there can be no escape.

The native American by the middle of the nineteenth century was rapidly becoming a distinct type. Derived from the Teutonic part of the British Isles, and being almost purely Nordic, he was on the point of developing physical peculiarities of his own, slightly variant from those of his English forefathers, and corresponding rather with the idealistic Elizabethan than with the materialistic Hanoverian Englishman. The Civil War, however, put a severe, perhaps fatal, check to the development and expansion of this splendid type, by de- , straying great numbers of the best breeding stock on both sides, and by breaking up the home ties of many more. If the war had not .occurred these same men with their descendants would have populated the Western States instead of the racial nondescripts who are now flocking there.

The prosperity that followed the war attracted hordes of newcomers who were welcomed by the native Americans to operate factories, build railroads, and fill up the waste spaces—"developing the country" it was called.

These new immigrants were no longer exclusively members of the Nordic race as were the earlier ones who came of their own impulse to improve social conditions. The transportation lines advertised America as a land flowing with milk and honey, and the European governments took the opportunity to unload upon careless, wealthy, and hospitable America the sweepings of their jails and asylums. The result was that the new immigration, while it still included many strong elements from the north of Europe, contained a large and increasing number of the weak, the broken, and the mentally crippled of all races drawn from the lowest stratum of the Mediterranean basin and the Balkans, together with hordes of the wretched, submerged populations of the Polish Ghettos.

With a pathetic and fatuous belief in the efficacy of American institutions and environment to reverse or obliterate immemorial hereditary tendencies, these newcomers were welcomed and given a share in our land and prosperity. The American taxed himself to sanitate and educate these poor helots, and as soon as they could speak English, encouraged them to enter into the political life, first of municipalities, and then of the nation.

The result is showing plainly in the rapid decline in the birth rate of native Americans because the poorer classes of Colonial stock, where they still exist, will not bring children into the world to compete in the labor market with the Slovak, the Italian, the Syrian, and the Jew. The native American is too proud to mix socially with them, and is gradually withdrawing from the scene, abandoning to these aliens the land which he conquered and developed. The man of the old stock is being crowded out of many country districts by these foreigners, just as he is to-day being literally driven off the streets of New York City by the swarms of Polish Jews. These immigrants adopt the language of the native American; they wear his clothes; they steal his name; and they are beginning to take his women, but they seldom adopt his religion or understand his ideals, and while he is being elbowed out of his own home the American looks calmly abroad and urges on others the suicidal ethics which are exterminating his own race.

As to what the future mixture will be it is evident that in large sections of the country the native American will entirely disappear. He will not intermarry with inferior races, and he cannot compete in the sweat shop and in the street trench with the newcomers. Large cities from the days of Rome, Alexandria, and Byzantium have always been gathering points of diverse races, but New York is becoming a *cloaca gentium* which will produce many amazing racial hybrids and some ethnic horrors that will be beyond the powers of future anthropologists to unravel.

One thing is certain: in any such mixture, the surviving traits will be determined by competition between the lowest and most primitive elements and the specialized traits of Nordic man; his stature, his light colored eyes, his fair skin and blond hair, his straight nose, and his splendid fighting and moral qualities, will have little part in the resultant mixture.

The "survival of the fittest" means the survival of the type best adapted to existing conditions of environment, to-day the tenement and factory, as in Colonial times they were the clearing of forests, fighting Indians, farming the fields, and sailing the Seven Seas. From the point of view of race it were better described as the "survival of the unfit."

This review of the colonies of Europe would be discouraging were it not that thus far little attention has been paid to the suitability of a new country for the particular colonists who migrate there. The process of sending out colonists is as old as mankind itself, and probably in the last analysis most of the chief races of the world, certainly most of the inhabitants of Europe, represent the descendants of successful colonists.

Success in colonization depends on the selection of new lands and climatic conditions in harmony with the immemorial requirements of the incoming race. The adjustment of each race to its own peculiar habitat is based on thousands of years of rigid selection which cannot be safely ignored. A certain isolation and freedom from competition with other races, for some centuries at least, is also important, so that the colonists may become habituated to their new surroundings.

# PART II

# EUROPEAN RACES IN HISTORY

## I
## EOLITHIC MAN

Before considering the living populations of Europe, we must give consideration to the extinct peoples that preceded them.

The science of anthropology is very recent—in its present form less than fifty years old—but it has already revolutionized our knowledge of the past and extended prehistory so that it is now measured not by thousands but by tens of thousands of years.

The history of man prior to the period of metals has been divided into ten or more subdivisions, many of them longer than the time covered by written records. Man has struggled up through the ages, to revert again and again into savagery and barbarism, but apparently retaining each time something gained by the travail of his ancestors.

So long as there is in the world a freely breeding stock or race that has in it an inherent capacity for development and growth, mankind will continue to ascend until, possibly through the selection and regulation of breeding as intelligently applied as in the case of domestic animals, he will control his own destiny and attain moral heights as yet unimagined.

The impulse upward, however, is supplied by a very small number of nations, and by a very small portion of the population in such nations. The section of any community that produces leaders or genius of any sort is only a minute percentage. To invent new processes, to establish new principles, to elucidate and unravel the laws of nature, calls for genius. To imitate or to adopt what others have invented is not genius but mimicry.

This something which we call "genius" is not a matter of family, but of stock or strain, and is inherited in precisely the same manner as are the purely physical characters. It may be latent through several generations of obscurity, and then flare up when the opportunity comes. Of this we have many examples in America. This is what education or opportunity does for a community; it permits in these rare cases fair play for development, but it is race, always race, that produces genius

This genius producing type is slow breeding, and there is real danger of its loss to mankind. Some idea of the value of these small strains can be gained from the recent statistics which demonstrate that Massachusetts produces more than fifty times as much genius per hundred thousand whites as does Georgia, Alabama, or Mississippi, although apparently the race, religion, and environment, other than climatic conditions, are much the same, except for the numbing presence in the South of a large stationary negro population.

The more thorough the study of European prehistory becomes, the more we realize how many advances of culture have been made and then lost. Our parents were accustomed to regard the overthrow of ancient civilization in the Dark Ages as the greatest catastrophe of mankind, but we now know that the classic period of Greece was preceded by similar dark ages caused by the Dorian invasions, which overthrew the Homeric-Mycenaean culture, which in its turn had flourished after the destruction of its parent, the Minoan culture of Crete. Still earlier, some twelve thousand years ago, the Azilian period of poverty and retrogression succeeded the wonderful achievements of the hunter-artists of the Upper Paleolithic.

The progress of civilization becomes evident only when immense periods are studied and compared, but the lesson is always the same, namely, that race is everything. Without race there can be nothing except the slave wearing his master's clothes, stealing his master's proud name, adopting his master's tongue, and living in the crumbling ruins of his master's palace. Everywhere on the sites of ancient civilization the Turk, the Kurd, and the Bedouin camp; and Americans might well pause and consider the fate of this country which they, and they alone, founded and nourished with their blood. The immigrant ditch diggers and the railroad navvies were to our fathers what their slaves were to the Romans, and the same transfer of political power from master to servant is taking place to-day.

Man's place of origin was undoubtedly Asia. Europe is only a peninsula of the Eurasiatic continent, and although the extent of its land area during the Pleistocene was much greater than at present, it is certain, from the distribution of the various species of man, that the main races evolved in Asia long before the centre of that continent was reduced to deserts by progressive desiccation.

Evidence of the location of the early evolution of man in Asia and the geologically recent submerged area toward the southeast is afforded by the fossil deposits in the Siwalik hills of northern India, where have been found the remains of primates which were either ancestral or closely related to the four genera of living anthropoids; and by the discovery in Java, which in Pliocene times was connected with the mainland over what is now the South China Sea, of the earliest known form of erect primate, the *Pithecanthropus*. This apelike man is practically the " missing link," being intermediate between man and the anthropoids.

Pithecanthropus is generally believed to have been contemporary with the Günz glaciation of some 500,000 years ago, the first of the four great glacial advances in Europe.

One or two forms of fossil anthropoid apes have been discovered in the Miocene of Europe which may possibly have been remotely related to the ancestors of man, but when the archaeological exploration of Asia shall be as complete and intensive as that of Europe, it is probable that more forms of fossil anthropoids and new species of man will be found there.

Man existed in Europe during the second and third interglacial periods, if not earlier. We have his artifacts in the form of eoliths, at least as early as the second interglacial stage, the Mindel-Riss, of some 300,000 years ago. A single jaw found near Heidelberg is referred to this period and is the earliest skeletal evidence of man in Europe. From certain remarkable characters in this jaw, it has been assigned to a new species, *Homo heidelbergensis*.

Then follows a long period of scanty industrial relics and no known skeletal remains. Man was slowly and painfully struggling up from an eolithic culture phase, where chance flints served his temporary'purpose. This in turn was succeeded by a stage of human development where slight chipping and retouching of flints for man's increasing needs led, after vast intervals of time, to the deliberate manufacture of tools. This period is known as the Eolithic, and is necessarily extremely hazy and uncertain. Whether or not certain chipped or broken flints, called eoliths, or dawn stones, were really human artifacts or were the products of natural forces is really immaterial because man must have passed through such an eolithic stage.

The further back we go toward the commencement of such an eolithic culture, the more and more unrecognizable the flints necessarily become until they finally cannot be distinguished from natural stone fragments, because at the beginning the earliest man merely picked up a convenient stone, used it once and flung it away, precisely as an anthropoid ape would act to-day if he wanted to break in the shell of a tortoise or crack an ostrich egg.

Man must have experienced the following phases of development in the transition from the prehuman to the human stage: first, the utilization of chance stones and sticks; second, the casual adaptation of flints by a minimum amount of chipping; third, the deliberate manufacture of the simplest implements from flint nodules; and fourth, the invention of new forms of weapons and tools in ever increasing variety.

Of the last two stages we have an extensive and clear record. Of the second stage we have in the eoliths intermediate forms ranging from flints that are evidently results of natural causes to flints that are clearly artifacts. The first and earliest stage, of course, could leave behind it no definite record and must always rest on hypothesis.

# II
# PALEOLITHIC MAN

With the deliberate manufacture of implements from flint nodules, we enter the beginning of Paleolithic time, and from here on our way is relatively clear. The successive stages of the Paleolithic were of great length, but are each characterized by some improvement in the manufacture of tools. During long ages man was merely a tool making and tool using animal, and, after all is said, that is about as good a definition as we can find to-day for the primate we call human.

The Paleolithic Period, or Old Stone Age, lasted from the somewhat indefinite termination of the Eolithic, some 150,000 years ago, to the Neolithic or New Stone Age, which began about 7,000 B. C.

The Paleolithic falls naturally into three great subdivisions. The Lower Paleolithic includes the whole of the last interglacial stage with the subdivisions of the Pre-Chellean, Chellean, and Acheulean; the Middle Paleolithic covers the whole of the last glaciation, and is co-extensive with the Mousterian Period and the dominance of the Neanderthal species of man. The Upper Paleolithic covers all the postglacial stages down to the Neolithic, and includes the subdivisions of the Aurignacian, Solutrean, Magdalenian, and Azilian. During the entire Upper Paleolithic, except the short closing phase, the Cro-Magnon race flourished.

It is not until after the third severe period of great cold, known as the Riss glaciation, and until we enter, some 150,000 years ago, the third and last interglacial stage of temperate climate, known as the Riss-Wurm, that we begin a definite and ascending series of culture. The Pre-

Chellean, Chellean and Acheulean divisions of the Lower Paleolithic occupied the whole of this warm or rather temperate interglacial phase, which lasted nearly 100,000 years.

A shattered skull, a jaw, and some teeth have been discovered recently in Sussex, England. These remains were all attributed to the same individual, who was named the Piltdown Man. Owing to the extraordinary thickness of the skull and the simian character of the jaw, a new genus, *Eoanthropus*, the "dawn man," was created and assigned to Pre-Chellean times. Further study and comparison with the jaws of other primates demonstrated that the jaw belonged to a chimpanzee, so that the genus *Eoanthropus* must now be abandoned, and the Piltdown Man must be included in the genus *Homo* as at present constituted. Future discoveries of the Piltdown type and for that matter of Heidelberg Man may, however, raise either or both of them to generic rank.

Some of the tentative restorations of the fragmentary bones make this skull altogether too modern and too capacious for a Pre-Chellean or even a Chellean. In any event the Piltdown Man is highly aberrant and, so far as our present knowledge goes, does not appear to be related to any other species of man found during the Lower Paleolithic.

In later, Acheulean, times a new species of man, very likely descended from the early Heidelberg Man of Eolithic times, appears on the scene, and is known as the Neanderthal race. Many fossil remains of this type have been found.

The Neanderthaloids occupied the European stage exclusively, with the possible exception of the Piltdown Man, so far as our information extends, from the first appearance of man in Europe to the end of the Middle Paleolithic. The Neanderthals flourished throughout the entire duration of the last glacial advance known as the Wurm glaciation. This period, known as the Mousterian, began about 50,000 years ago, and lasted some 25,000 years.

The Neanderthal species disappears suddenly and completely with the advent of postglacial times, when, about 25,000 years ago, he was apparently exterminated by a new and far higher race, the famous Cro-Magnons.

There may well have been, and probably were, during Mousterian times, races of man in Europe other than the Neanderthaloids, but of them we have no record. Among the numerous remains of Neanderthals, however, we do find traces of distinct types showing that this race in Europe was undergoing evolution and was developing marked variations in characters.

Neanderthal Man was a purely meat eating hunter, living in caves, or rather in their entrances. He was dolichocephalic and not unlike existing Australoids, although not necessarily of black skin, and was, of course, in no sense a negro.

The skull was characterized by heavy superorbital ridges, a low, receding forehead, protruding and chinless under jaw, and the posture was imperfectly erect. This race was widely spread and rather numerous. Some of its blood has trickled down to the present time, and occasionally one sees a skull of the Neanderthal type. The best skull of this type ever seen by the writer belonged to an old and very intellectual professor in London, who was quite innocent of his value as a museum specimen. In the old black breed of Scotland the overhanging brow and deep-set eyes are suggestive of this race.

Along with other ancient and primitive racial remnants, ferocious gorilla like living specimens of the Neanderthal man are found not infrequently on the west coast of Ireland, and are easily recognized by the great upper lip, bridgeless nose, beetling brow and low growing hair, and wild and savage aspect. The proportions of the skull which give rise to this large upper lip, the low forehead, and the superorbital ridges are clearly Neanderthal characters. The other traits of this Irish type are common to many primitive races. This is the Irishman of caricature, and the type was very frequent in America when the first Irish immigrants came in 1846 and the following years. It seems, however, to have almost disappeared in this country.

In the Upper Paleolithic, which began after the close of the fourth and last glaciation, about 25,000 years ago, the Neanderthal race was succeeded by men of very modern aspect, known as Cro-Magnons. The date of the beginning of the Upper Paleolithic is the first we can fix with accuracy, and its correctness can be relied on within narrow limits. The Cro-Magnon race first appears in the Aurignacian subdivision of the Upper Paleolithic. Like the Neanderthals, they were dolichocephalic, with a cranial capacity superior to the average in existing European populations, and a stature of very remarkable size.

It is quite astonishing to find that the predominant race in Europe 25,000 years ago, or more, was not only much taller, but had an absolute cranial capacity in excess of the average of the present population. The low cranial average of existing populations in Europe can be best explained by the presence of large numbers of individuals of inferior mentality. These defectives have been carefully preserved by modern charity, whereas in the savage state of society the backward members are allowed to perish and the race is carried on by the vigorous and not by the weaklings.

The high brain capacity of the Cro-Magnons is paralleled by that of the ancient Greeks, who in a single century gave to the world out of their small population very much more genius than all the other races of mankind have since succeeded in producing in a similar length of time. Athens between 530 and 430 B. C. had an average population of about 90,000 freemen, and yet from these small numbers there were born no less than fourteen geniuses of the very highest rank. This would indicate a general intellectual status as much above that of the Anglo-Saxons as the latter are above the negroes. The existence at these early dates of a very high cranial capacity and its later decline shows that there is no upward tendency inherent in mankind of sufficient strength to overcome obstacles placed in its way by stupid social customs.

All historians are familiar with the phenomenon of a rise and decline in civilization such as has occurred time and again in the history of the world, but we have here in the disappearance of the Cro-Magnon race the earliest example of the replacement of a very superior race by an inferior one. There is great danger of a similar replacement of a higher by a lower type here in America unless the native American uses his superior intelligence to protect himself and his children from competition with intrusive peoples drained from the lowest races of eastern Europe and western Asia.

While the skull of the Cro-Magnon was long, the cheek bones were very broad, and this combination of broad face with long skull constitutes a peculiar disharmonic type which occurs to-day only among the very highly specialized Esquimaux and one or two other unimportant groups.

Skulls of this particular type, however, are found in small numbers among existing populations in central France, precisely in the district where the fossil remains of this race were first discovered. These isolated Frenchmen probably represent the last lingering remnant of this splendid race of hunting savages.

The Cro-Magnon culture is found all around the basin of the Mediterranean, and this fact, together with the conspicuous absence in eastern Europe of its earliest phases, the lower Aurignacian, indicates that it entered Europe by way of north Africa, precisely as did, in Neolithic times, its successors, the Mediterranean race. There is little doubt that the Cro-Magnons originally developed in Asia and were in their highest stage of physical development at the time of their first appearance in Europe. Whatever change took place in their stature during their residence there seems to have been in the nature of a decline rather than of a further development.

There is nothing whatever of the negroid in the Cro-Magnons, and they are not in any way related to the Neanderthals, who represent a distinct and extinct species of man.

The Cro-Magnon race persisted through the entire Upper Paleolithic, during the periods known as the Aurignacian, Solutrean, and Magdalenian, from 25,000 to 10,000 B. C. While it is possible that the blood of this race enters somewhat into the composition of the peoples of western Europe, its influence cannot be great, and the Cro-Magnons disappear from view with the advent of the warmer climate of recent times.

It has been suggested that, following the fading ice edge north and eastward through Asia into North America, they became the ancestors of the Esquimaux, but certain anatomical objections are fatal to this interesting theory. No one, however, who is familiar with the culture of the Esquimaux, and especially with their wonderful skill in bone-carving, can fail to be struck with the similarity of their technique to that of the Cro-Magnons.

To the Cro-Magnon race the world owes the birth of art. Caverns and shelters are yearly uncovered in France and Spain, where the walls and ceilings are covered with polychrome paintings or with incised bas-reliefs of animals of the chase. A few clay models, sometimes of the human form, are also found together with abundant remains of their chipped but unpolished stone weapons and tools. Certain facts stand out clearly, namely, that they were pure hunters and clothed themselves in furs and skins. They knew nothing of agriculture or of domestic animals, even the dog being as yet untamed, and the horse was regarded merely as an object of chase.

The question of their knowledge of the principle of the bow and arrow during the Aurignacian and Solutrean is an open one, but there are definite indications of the use of the arrow, or at least the barbed dart, in early Magdalenian times, and this weapon was well known in the succeeding Azilian Period.

The presence toward the end of this last period of quantities of very small flints, called microliths, has given rise to much controversy. It is possible that these microliths represent the tips of small poisoned arrows such as are now in very general use among primitive hunting tribes the world over. Certain grooves in some of the flint weapons of the Upper Paleolithic may well have been also used for the reception of poison. It is highly probable that these skilful savages, the Cro-Magnons, perhaps the greatest hunters that ever lived, not only used poisoned darts, but were adepts in trapping game by means of pitfalls and snares, precisely as do some of the hunting tribes of Africa to-day. Barbed arrowheads of flint or bone, such as were commonly used by the North American Indians, have not been found in Paleolithic deposits.

In the next period, the Solutrean, the Cro-Magnons shared Europe with a new race known as the Brünn-Předmost, found in central Europe. This race is characterized by a long face as well as a long skull, and was, therefore, harmonic. This Brünn-Předmost race would appear to have been well settled in the Danubian and Hungarian plains, and this location indicates an eastern rather than a southern origin.

Good anatomists have seen in this race the last lingering traces of the Neanderthaloids, but it is more probable that we have here the first advance wave of the primitive forerunners of one of the modern European dolichocephalic races.

This new race was not artistic, but had great skill in fashioning weapons. It is possibly associated with the peculiarities of Solutrean culture and the decline of art which characterizes that period. The artistic impulse of the Cro-Magnons which flourished so vigorously during the Aurignacian, seems to be quite suspended during this Solutrean period, but reappears in the succeeding Magdalenian times. This Magdalenian art is clearly the direct descendant of Aurignacian models, and in this closing age of the Cro-Magnons all forms of Paleolithic art, carving, engraving, painting, and the manufacture of weapons, reach their highest and final culmination.

Nine thousand or ten thousand years may be assigned for the Aurignacian and Solutrean Periods, and we may with considerable certainty give the minimum date of 16,000 B. C. for the beginning of Magdalenian time. Its entire duration can be safely set down at 6,000 years, thus bringing the final termination of the Magdalenian to 10,000 B. C. All these dates are extremely conservative, and the error, if any, would be in assigning too late and not too early a period to the end of Magdalenian times.

At the close of the Magdalenian we enter upon the last period of Paleolithic times, the Azilian, which lasted from about 10,000 to 7,000 B. C., when the Upper Paleolithic, the age of chipped flints, definitely and finally ends. This period takes its name from the Mas d'Azil or "House of Refuge," a huge cavern in the eastern Pyrenees, where the local Protestants took shelter during the persecutions. In this cave the extensive deposits are typical of this, epoch, and here certain marked pebbles show the earliest known traces of the alphabet.

With the advent of this closing Azilian Period art entirely disappears, and the splendid physical specimens of the Cro-Magnons are succeeded by what appear to have been degraded savages, who had lost the force and vigor necessary for the strenuous chase of larger game, and had turned to the easier life of fishermen.

The bow and arrow in the Azilian are in common use in Spain, and it is well within the possibilities that the introduction of this new weapon from the south may have played its part in the destruction of the Cro-Magnons; otherwise it is hard to account for the disappearance of this race of large stature and great brain power.

The Azilian, also called the Tardenoisian in the north of France, was evidently a period of racial disturbance, and at its close the beginnings of the existing races are found.

From the first appearance of man in Europe, and for many tens of thousands of years down to some ten or twelve thousand years ago, all known human remains are of dolichocephalic type.

In the Azilian Period there appears the first round skull race. It comes clearly from the east. Later we shall find that this invasion of the fore-runners of the existing Alpine race came from southwestern Asia by way of the Iranian plateaux, Asia Minor, the Balkans, and the valley of the Danube, and spread over nearly all of Europe. The earlier round skull invasions may as well have been infiltrations as armed conquests, since apparently from that day to this the round skulls have occupied the poorer mountain districts and have seldom ventured down to the rich and fertile plains.

This new brachycephalic race is known as the Furfooz or Grenelle race, so called from the localities in Belgium and France where it was first discovered. Members of this round skull race have also been found at Of net, in Bavaria, where they occur in association with a dolichocephalic race, our first historic evidence of the mixture of contrasted races. The descendants of this Furfooz-Grenelle race and of the succeeding waves of invaders of the same brachycephalic type now occupy central Europe as Alpines and form the predominant peasant type in central and eastern Europe.

In this same Azilian Period there appear, coming this time from the south, the first forerunners of the Mediterranean race. The descendants of this earliest wave of Mediterraneans and their later reinforcements occupy all the coast and islands of the Mediterranean, and are spread widely over western Europe. They can everywhere be identified by their short stature, long skull, and brunet hair and eyes.

While during this Azilian-Tardenoisian Period these ancestors of two of the existing European races are appearing in-central and southern Europe, a new culture phase, also distinctly Pre-Neolithic, was developing along the shores of the Baltic. It is known as Maglemose from its type locality in Denmark. It is probably the work of the first wave of the Nordic subspecies, possibly the Proto-Teutons, who had followed the retreating glaciers north over the old land connections between Denmark and Sweden to occupy the Scandinavian Peninsula. In the remains of this culture we find for the first time definite evidence of the domesticated dog. As yet, however, no skeletal remains have been discovered.

With the appearance of the Mediterranean race the Azilian-Tardenoisian draws to its close, and with it the entire Paleolithic Period. It is safe to assign for the end of the Paleolithic and the beginning of the Neolithic or Polished Stone Age, the date of 7000 or 8000 B. C.

The races of the Paleolithic Period arrived successively on the scene with all their characters fully developed. The evolution of all these subspecies and races took place somewhere in Asia or eastern Europe. None of these races appear to be ancestral one to another, although the scanty remains of the Heidelberg Man would indicate that he may have given rise to the later Neanderthals. Other than this possible affinity, the various races of Paleolithic times are not related one to another.

# III

# THE NEOLITHIC AND BRONZE AGES

About 7000 B. C. we enter an entirely new period in the history of man, the Neolithic or New Stone Age, when the flint implements were polished and not merely chipped. Early as is this date in European culture, we are not far from the beginnings of an elaborate civilization in parts of Asia. The earliest organized states, so far as our present knowledge goes, were the Mesopotamian empires of Accad and Sumer—though they may have been preceded by the Chinese civilization, whose origin remains a mystery, nor can we trace any connection between it and western Asia. Balkh, the ancient Bactra, the mother of cities, is located where the trade routes between China, India, and Mesopotamia converged, and it is in this neighborhood that careful and thorough excavations will probably find their greatest rewards.

However, we are not dealing with Asia, but with Europe only, and our knowledge is confined to the fact that the various cultural advances at the end of the Paleolithic and the beginning of the Neolithic correspond with the arrival of new races.

The transition from the Paleolithic to the Neolithic was formerly considered as revolutionary, an abrupt change of both race and culture, but a period more or less transitory, known as the Campignian, now appears to bridge over this gap. This is but what should be expected, since in human archaeology as in geology the more detailed our knowledge becomes, the more gradually we find one period or horizon merges into its successor.

For a long time after the opening of the Neolithic the old fashioned chipped weapons and implements remain the predominant type, and the polished flints so characteristic of the Neolithic appear at first only sporadically, then increase in number, until finally they entirely replace the rougher designs of the preceding Old Stone Age.

So in turn these Neolithic polished stone implements which ultimately became both varied and effective as weapons and tools, continued in use long after metallurgy developed. In the Bronze Period, of course, metal armor and weapons were for ages of the greatest value. So they were necessarily in the possession of the military and ruling classes only, while the unfortunate serf or common soldier who followed his master to war did the best he could with leather shield and stone weapons. In the ring that clustered around Harold for the last stand on Senlac Hill many of the English thanes died with their Saxon king, armed solely with the stone battle-axes of their ancestors.

In Italy also there was a long period known to the Italian archaeologists as the Eneolithic Period, when good flint tools existed side by side with very poor copper and bronze implements; so that, while the Neolithic lasted in western Europe four or five thousand years, it is, at its commencement, without clear definition from the preceding Paleolithic, and at its end it merged gradually into the succeeding ages of metals.

After the opening Campignian phase there followed a long period typical of the Neolithic, known as the Robenhausian, or Age of the Swiss Lake Dwellers, which reached its height about 5000 B. C. The lake dwellings seem to have been the work exclusively of the round skull Alpine races and are found in numbers throughout the region of the Alps and their foothills and along the Danube valley.

These Robenhausian pile built villages were in Europe the earliest known form of fixed habitation, and the culture found in association with them was a great advance on that of the preceding Paleolithic. This type of permanent habitation flourished through the entire Upper Neolithic and the succeeding Bronze Age. Pile villages end in Switzerland with the first appearance of iron, but elsewhere, as in the upper Danube, they still existed in the days of Herodotus.

Domesticated animals and agriculture, as well as rough pottery, appear during the Robenhausian for the first time. The chase, supplemented by trapping and fishing, was still common, but it probably was more for clothing than for food. Of course, a permanent site is the basis of an agricultural community, and involves at least a partial abandonment of the chase, because only nomads can follow the game in its seasonal migrations, and hunted animals soon leave the neighborhood of settlements.

The Terramara Period of northern Italy was a later phase of culture contemporaneous with the Upper Robenhausian, and was typical of the Bronze Age. During the Terramara Period fortified and moated stations in swamps or close to the banks of rivers became the favorite resorts instead of pile villages built in lakes. The first traces of copper are found during this period. The earliest human remains in the Terramara deposits are long skulled, but round skulls soon appear in association with bronze implements. This indicates an original population of Mediterranean affinities swamped later by Alpines.

Neolithic culture also flourished in the north of Europe and particularly in Scandinavia, now free from ice. The coasts of the Baltic were apparently occupied for the first time at the very beginning of this period, as no trace of Paleolithic industry has been found there,. other than the Maglemose, which represents only the very latest phase of the Old Stone Age. The kitchen middens, or refuse heaps, of Sweden, and more particularly of Denmark, date from the early Neolithic, and thus are somewhat earlier than the lake dwellers. No trace of agriculture has been found in them, and the dog seems to have been the only domesticated animal.

From these two centres, the Alps and the North, an elaborate and variegated Neolithic culture spread through western Europe, and an autochthonous development took place little influenced by trade intercourse with Asia after the first immigrations of the new races.

We may assume that the distribution of races during the Neolithic was roughly as follows: The Mediterranean basin and western Europe, including Spain, Italy, Gaul, Britain, and the western portions of Germany, populated by Mediterranean long heads; the Alps and the territories immediately surrounding, except the valley of the Po, together with much of the Balkans, inhabited by Alpine types. These Alpines extended northward until they came in touch in eastern Germany and Poland with the southernmost Nordics, but as the Carpathians at a much later date, namely from the fourth to the eighth century A. D., were the centre of radiation of the Alpine Slavs, it is very possible that during the Neolithic the early Nordics lay farther north and east.

North of the Alpines and occupying the shores of the Baltic and Scandinavia, together with eastern Germany, Poland, and Russia, were located the Nordics. At the very base of the Neolithic, and perhaps still earlier, this race occupied Scandinavia, and Sweden became the nursery of the Teutonic subdivision of the Nordic race. It was in that country that the peculiar characters of stature and blondness became most accentuated, and it is there that we find them to-day in their greatest purity. During the Neolithic the remnants of early Paleolithic man must have been numerous, but later they were either exterminated or absorbed by the existing European races.

During all this Neolithic Period Mesopotamia and Egypt were thousands of years in advance of Europe, but only a small amount of culture from these sources seems to have trickled westward up the valley of the Danube, then and long afterward the main route of intercourse between western Asia and the heart of Europe. Some trade also passed from the Black Sea up the Russian rivers to the Baltic coasts. Along these latter routes there came from the north to the Mediterranean world the amber of the Baltic, a fossil resin greatly prized by early man for its magic electrical qualities.

Gold was probably the first metal to attract the attention of primitive man, but, of course, could only be used for purposes of ornamentation. Copper, which is often found in a pure state, was also one of the earliest metals known, and probably came first either from the mines of Cyprus or of the Sinai Peninsula. These latter mines are known to have been worked before 3800 B. C. by systematic mining operations, and much earlier the metal must have been obtained by primitive methods from surface ore. It is, therefore, probable that copper was known and used, at first for ornament and later for implements, in Egypt before 5000 B. C., and probably even earlier in the Mesopotamian regions.

With the use of copper the Neolithic fades to its end and the Bronze Age commences soon thereafter. This next step in advance was made apparently about 4000 B. C., when some unknown genius discovered that an amalgam of nine parts of copper to one part of tin would produce the metal we now call bronze, which has a texture and strength suitable for weapons and tools. The discovery revolutionized the world. The new knowledge was a long time spreading and weapons of this material were of fabulous value, especially in countries where there were no native mines, and where spears and swords could only be obtained through trade or conquest. The esteem in which these bronze weapons, and still more the later weapons of iron, were held, is indicated by the innumerable legends and myths concerning magic swords and armor, the possession of which made the owner well-nigh invulnerable and invincible.

The necessity of obtaining tin for this amalgam led to the early voyages of the Phoenicians, who from the cities of Tyre and Sidon, and their daughter, Carthage, traversed the entire length of the Mediterranean, founded colonies in Spain to work the Spanish tin mines, passed the Pillars of Hercules, and finally voyaged through the stormy Atlantic to the Cassiterides, the Tin Isles of Ultima Thule. There, on the coasts of Cornwall, they traded with the native British, of kindred Mediterranean race, for the precious tin. These dangerous and costly voyages become explicable only if the value of this metal for the composition of bronze be taken into consideration.

After these bronze weapons were elaborated in Egypt, the knowledge of their manufacture and use was extended through conquest into Palestine, and about 3000 B. C. northward into Asia Minor.

The effect of the possession of these new weapons on the Alpine populations of western Asia was magical, and resulted in an intensive and final expansion of round skulls into Europe. This invasion came through Asia Minor, the Balkans, and the valley of the Danube, poured into Italy from the north, introduced bronze among the earlier Alpine lake dwellers of Switzerland, and among the Mediterraneans of the Terramara stations of the valley of the Po, and at a later date reached as far west as Britain and as far north as Holland and Norway.

The simultaneous appearance of bronze about 3000 or 2800 B. C. in the south as well as in the north of Italy can probably be attributed to a wave of this same invasion which reached Tunis and Sicily, passing through Egypt, where it left behind the so-called Giza round skulls. With the first knowledge of metals begins the Eneolithic Period of the Italians.

The introduction into England and into Scandinavia of bronze may be safely dated about one thousand years later, around 1800 B. C. The fact that the Alpines only barely reached Ireland, and that the invasion of Britain itself was not sufficiently intensive to leave any substantial record of its passing in the skulls of the existing population, indicates that at this time Ireland was severed from England, and that the land connection between England and France had been broken. The computation of the foregoing dates, of course, is somewhat hypothetical, but the fixed fact remains that this last expansion of the Alpines brought the knowledge of bronze to western and northern Europe and to the Mediterranean and Nordic peoples living there.

## CHRONOLOGICAL TABLE *

| Period | Culture / Region | Race | Date |
|---|---|---|---|
| **METALS** | | | |
| LATER IRON — La Tène Culture | Europe | | 500 B. C.–Roman times |
| EARLY IRON — Hallstatt Culture | Europe | | 1500–500 B. C. |
| | Orient | | 1800–1000 B. C. |
| BRONZE | Western and northern Europe | | 1800–500 B. C. |
| | Orient | | 4000–2000 B. C. |
| **NEOLITHIC** | | | |
| LATE NEOLITHIC, COPPER, ENEOLITHIC | | | 3000–2000 B. C. |
| TYPICAL NEOLITHIC | Swiss lake dwellings, Robenhausian culture | | 5000 B. C. |
| EARLY NEOLITHIC | Campignian culture | | 7000 B. C. |
| **UPPER PALEOLITHIC** | | | |
| POSTGLACIAL (Caves and shelters) | Azilian-Tardenoisian | Furfooz-Grenelle race; Proto-Mediterranean race | 10,000–7000 B. C. |
| | Magdalenian | Cro-Magnon race | 16,000–10,000 B. C. |
| | Solutrean | Brünn-Předmost race; Cro-Magnon race | 25,000–16,000 B. C. |
| | Aurignacian | Cro-Magnon race | 25,000–16,000 B. C. |

|  |  |  |  |
|---|---|---|---|
| **MIDDLE PALEOLITHIC** | | | |
| IV. GLACIATION | Würm | Mousterian { Neanderthal race; Caves and shelters } | 50,000–25,000 B. C. |
| **LOWER PALEOLITHIC** | | | |
| III. INTERGLACIAL | Riss-Würm | Acheulean, river terraces | 75,000 B. C. |
| | | Chellean, river terraces | 100,000 B. C. |
| | | Pre-Chellean and Mesvinian, river terraces | 125,000 B. C. / 150,000 B. C. |
| **EOLITHIC** | | | |
| III. GLACIATION | Riss | | 200,000–150,000 B. C. |
| II. INTERGLACIAL | Mindel-Riss | Heidelberg Man | 350,000–200,000 B. C. |
| II. GLACIATION | Mindel | | 400,000–350,000 B. C. |
| I. INTERGLACIAL | Günz-Mindel | | 475,000–400,000 B. C. |
| FIRST GLACIAL | Günz | *Pithecanthropus* | 500,000–475,000 B. C. |

* After Henry Fairfield Osborn, 1915.

The effect of the introduction of bronze in the areas occupied chiefly by the Mediterranean race along the Atlantic coast and in Britain, as well as in North Africa from Tunis to Morocco, is seen in the wide distribution of the megalithic funeral monuments, which appear to have been erected, not by Alpines, but by the dolichocephs. The occurrence of bronze tools and weapons in the interments shows clearly that the megaliths date from this Bronze Age. But their construction and use continued at least until the very earliest trace of iron appeared, and in fact mound burials among the Vikings were common until the introduction of Christianity.

The knowledge of iron as well as bronze in Europe, centres around the area occupied by the Alpines in the eastern Alps and its earliest phase is known as the Hallstatt culture, from a little town in the Tyrol where it was first discovered. This Hallstatt iron culture flourished about 1500 B. C. Whether or not the Alpines introduced from Asia or invented in Europe the smelting of iron, it was the Nordics who benefited by its use. Bronze weapons and the later iron ones proved in the hands of these northern barbarians to be of terrible effectiveness, and were first of all turned against their Alpine teachers. With these metal swords in their grasp, the Nordics first conquered the Alpines of central Europe and then suddenly entered the ancient world as raiders and destroyers of cities, and the classic civilizations of the north coasts of the Mediterranean Sea fell, one after another, before the "Furor Normanorum," just as two thousand years later the provinces of Rome were devastated by the last wave of the men of the north, the Teutonic tribes.

The first Nordics to appear in European history are tribes speaking Aryan tongues, in the form of the various Celtic and related dialects in the west, of Umbrian in Italy and of Thracian in the Balkans, and these tribes, pouring down from the north, swept with them large numbers of Alpines, whom they had already thoroughly Nordicized. The process of conquering and assimilating these Alpines must have gone on for long centuries before our first historic records, and the work was so thoroughly done that the very existence of this Alpine race as a separate subspecies of man was actually forgotten for thousands of years by themselves and by the world at large, until it was revealed in our own day by the science of skull measurements.

The Hallstatt iron culture did not extend into western Europe, and the smelting and extensive use of iron in south Britain and northwest Europe are of much later date and occur in what is known as the La Tène Period, usually assigned to the fifth and fourth century B. C. Iron weapons were known in England much earlier, perhaps as far back as 800 or 1000 B.

C., but were very rare and were probably importations from the Continent.

The spread of this La Tène culture is associated with the Cymry, who constituted the last wave of Celtic-speaking invaders into western Europe, while the earlier Nordic Gauls and Goidels had arrived in Gaul and Britain equipped with bronze only.

In Roman times, which follow the La Tène Period, the three main races of Europe occupied the relative positions which they had held during the whole Neolithic Period and which they hold today, with the exception that the Nordic species was less extensively represented in western Europe than when, a few hundred years later, the Teutonic tribes flooded these countries; but on the other hand, the Nordics occupied large areas in eastern Germany, Hungary, Poland, and Russia now occupied by the Slavs of Alpine race, and many countries also in central Europe were in Roman times inhabited by fair haired, blue eyed barbarians, where now the population is preponderantly brunet and becoming yearly more so.

# IV
# THE ALPINE RACE

The Alpine race is clearly of Eastern and Asiatic origin. It forms the westernmost extension of a widespread subspecies which, outside of Europe, occupies Asia Minor, Iran, the Pamirs, and the Hindu Kush. In fact the western Himalayas were probably its centre of original evolution and radiation, and its Asiatic members constitute a distinct subdivision, the Armenoids.

The Alpine race is distinguished by a round face and correspondingly round skull which in the true Armenians has a peculiar, sugarloaf shape, a character which can be easily recognized. The Alpines must not be confounded with the sliteyed Mongols who centre around Thibet and the steppes of north Asia. The fact that both these races are round skulled does not involve identity of origin any more than the long skulls of the Nordics and of the Mediterraneans require that they be both considered of the same subspecies, although good anthropologists have been misled by this parallelism. The Alpines are of stocky build and moderately short stature, except where they have been crossed with Nordic elements. This race is also characterized by dark hair, tending to a dark brown color, and in Europe at the present time the eye is usually dark but sometimes grayish. The ancestral Proto-Alpines from the highlands of western Asia must, of course, have had brunet eyes, and very dark, probably black, hair. Whether we are justified in considering gray eyes peculiar to populations of mixed Alpine and Nordic blood is difficult to determine, but one thing is certain, the combination of blue eyes and flaxen hair is never Alpine.

The European Alpines retain very little of their Asiatic origin, except the skull, and have been in contact with the Nordic race so long that in central and western Europe they are everywhere saturated with the blood of that race. Many populations now considered good Germans, such as the majority of the Würtembergers, Bavarians,

Austrians, Swiss, and Tyrolese, are merely Teutonized Alpines.

The first appearance in Europe of the Alpines, dates from the Azilian Period when it is represented by the Furfooz-Grenelle race. There were, later, several invasions of this race which entered Europe during Neolithic times from the Asia Minor plateaux, by way of the Balkans and the valley of the Danube. It appears also to have passed north of the Black Sea, as some slight traces have been discovered there of round skulls which long ante-date the existing population, but the Russian brachycephaly of to-day is of much later origin.

This race in its final expansion far to the northwest, ultimately reached Norway, Denmark, and Holland, and planted among the dolichocephalic natives small colonies of round skulls, which still exist. When this invasion reached the extreme northwest of Europe its energy was spent, and the invaders were soon forced back into central Europe by the Nordics. The Alpines at this time of maximum extension, about 1800 B. C., crossed into Britain, and a few reached Ireland and introduced bronze into both these islands. As the metal appears about the same time in Sweden, it is safe to assume that it was introduced by this same invasion, a record of which persists to this day in the existence of a colony of round skulls in southwest Norway.

Bronze culture everywhere antedates the earliest appearance of the Celtic-speaking Nordics in western Europe.

The men of the Round Barrows in England were Alpines, but their numbers were so scanty that they have not left behind them in the skulls of the living population any demonstrable evidence of their conquest. If we are ever able to accurately dissect out the various strains that enter, in more or less minute quantities, into the blood of the British Isles, we shall find traces of these Round Barrow men as well as other interesting and ancient remnants, especially in the western isles and peninsulas.

In the study of European populations the great and fundamental fact about the British Isles is the absence there to-day of Alpine round skulls.

It is the only important state in Europe in which the round skulls play no part, and the only nation of any rank composed solely of Nordic and Mediterranean races in approximately equal numbers. To this fact is undoubtedly due many of the individualities of the English nation.

The invasion of central Europe by Alpines, which occurred in the Neolithic, following in the wake of the Azilian forerunners of the same type—the Furfooz-Grenelle race—represented a very great advance in culture. They brought with them from Asia the art of domesticating animals and the first knowledge of the cereals and of pottery, and were an agricultural race in sharp contrast to the flesh eating hunters who preceded them.

The Neolithic populations of the lake dwellings in Switzerland and the extreme north of Italy, which flourished about 5000 B. C., all belonged to this Alpine race. A comparison of the scanty physical remains of these lake dwellers with the inhabitants of the existing villages on the lake shores demonstrates that the skull shape has changed little or not at all during the last seven thousand years, and affords us another proof of the persistency of unit characters.

This Alpine race in Europe is now so thoroughly acclimated that it is no longer Asiatic in any respect, and has nothing in common with the Mongols except its round skulls. Such Mongolian elements as exist today in scattered groups throughout eastern Europe are remnants of the later invasions of Tatar hordes which, beginning with Attila in the fifth century, ravaged eastern Europe for hundreds of years.

In western and central Europe the present distribution of the Alpine race is a substantial recession from its original extent, and it has been everywhere conquered and completely swamped by Celtic and Teutonic speaking Nordics. Beginning with the first appearance of the Celtic-speaking Nordics in western Europe, this race has been obliged to give ground, but has mingled its blood everywhere with the conquerors, and now after centuries of obscurity it appears to be increasing again at the expense of the master race.

The Alpines reached Spain, as they reached Britain, in small numbers and with spent force, but they still exist along the Cantabrian Alps as well as on the northern side of the Pyrenees, among the French Basques. There are also dim traces all along the north African coast of a round skull invasion about 3000 B. C. through Syria, Egypt, Tripoli, and Tunis, and from there through Sicily to southern Italy.

THEIR CHARACTERS AND

| European Races | Modern Peoples | Ancient Peoples | Skull Cephalic Ini |
|---|---|---|---|
| *Nordic.* Homo sapiens europeus, Baltic, Scandinavian, Teutonic, Germanic, Dolicho-lepto, Reihengraber, Finnic. | All Norse, Swedes, Danes, Letts, many Finlanders, many Russians and Poles, North Germans, many French, Dutch, Flemings, English, Scotch, most Irish, Native Americans, Canadians, Australians, Africanders. | Sacæ, Kassites, Cimmerians, Persians, Phrygians, Achæans, Dorians, Thracians, Umbrians, Oscans, Scythians, Gauls, Galatians, Cymry, Belgæ, many Romans, Goths, Lombards, Vandals, Burgunds, Franks, Danes, Saxons, Angles, Norse, Normans, Varangians. Reihengraber. Kurgans. Maglemose culture. | Long. 79 and less |
| *Alpine.* Homo sapiens alpinus, Celto-Slav of the French, Sarmatian, Arvernian, Slavic, Lappanoid, Armenoid. | Bretons, Walloons, Central French, some Basques, Savoyards, Swiss, Tyrolese, most South Germans, North Italians, German-Austrians, Magyars, many Poles, most Russians, Serbs, Bulgars, most Rumanians, most Greeks, Turks, Armenians, most Persians and Afghans. | Accadians, Sumerians, Hittites, Medes. Giza skulls, Swiss Lake Dwellers, Furfooz-Grenelle race. Robenhausen. Round Barrows. Bronze culture. | Round. 80 and ov |
| *Mediterranean.* Homo sapiens mediterraneus, Iberian, Ligurian, Atlanto-Med. | Many English, Portuguese, Spaniards, some Basques, Provençals, South Italians, Sicilians, many Greeks and Rumanians, Moors, Berbers, Egyptians, Kurds, many Persians and Afghans, Hindus. | Pelasgians, Etruscans, Ligurians, Phœnicians, most Greeks, many Romans, Cretans, Iberians. Long Barrows. Neolithic culture. | Long. 79 and less |
| *Upper Paleolithic. Extinct races.* Furfooz-Grenelle. | | Proto-Alpines. | Round, 79– |
| Brünn Předmost. | | | Long, 66– |
| Homo sapiens cromagnonensis. | A few Dordognois. | Cro-Magnons. | Long, with disharmo broad fac 63–76. |
| *Middle Paleolithic.* Homo neanderthalensis, Homo primigenius. | Traces among west Irish and among the old black breed of Scotland. | Neanderthals. Neanderthaloids. | Long. |

STRIBUTION

| Face | Nose | Stature | Hair Color | Eye Color | Language |
|---|---|---|---|---|---|
| High.<br>Narrow.<br>Long. | Narrow.<br>Straight.<br>Aquiline. | Tall. | Flaxen.<br>Fair.<br>Red.<br>Light brown to chestnut.<br>Never black. | Blue.<br>Gray.<br>Green. | All Aryan except Tchouds, Esths, many Finlanders, and a few tribes in Siberia. |
| Broad. | Variable.<br>Rather broad.<br>Coarse. | Medium.<br>Stocky.<br>Heavy. | Dark brown.<br>Black. | Black or dark brown.<br>Often hazel or gray, in western Europe. | In Europe all Aryan except Magyars and some Basques.<br>In Asia mostly Aryan, except Turcomans, Kirghizes, and other nomad tribes. |
| High.<br>Narrow.<br>Long. | Rather broad. | Short.<br>Slender. | Dark brown.<br>Black. | Black.<br>Dark brown. | In Europe all Aryan, except some Basques. In Africa all non-Aryan. In Asia all Aryan, except Dravidians and other Indian tribes. |
| Medium. | | | Probably very dark. | Probably very dark. | Non-Aryan. |
| Low and medium. | | | | | |
| Low and broad. | Narrow and aquiline. | Very tall and medium. | Probably very dark. | Probably very dark. | Non-Aryan. |
| Long. | Broad. | Short and powerful. | Probably very dark. | Probably very dark. | Non-Aryan. |

The Alpine race forms to-day, as in Caesar's time, the great bulk of the population of central France, with a Nordic aristocracy resting upon it. They occupy, as the lower classes, the uplands' of Belgium, where, known as Walloons, they speak an archaic French dialect closely related to the ancient *langue d'oil*. They form a majority of the upland population of Alsace, Lorraine, Baden, Würtemberg, Bavaria, Tyrol, Switzerland, and north Italy; in short of the entire central massif of Europe. In Bavaria and the Tyrol the Alpines are so thoroughly Teutonized that their true racial affinities are betrayed by their round skulls alone.

When we reach Austria we come in contact with the Slavic-speaking nations which form a subdivision of the Alpine race, appearing late in history and radiating from the Carpathian Mountains. In western and central Europe, in relation to the Nordic race, the Alpine is everywhere the ancient, underlying, and submerged type. The fertile lands, river valleys, and the cities are in the hands of the Teutons, but in eastern Germany and Poland we fin d conditions reversed. Here is an old Nordic broodland, with a Nordic substratum underlying the bulk of the peasantry, which now consists of round skull Alpine Slavs. On top of these again we have an aristocratic upper class of relatively recent introduction. In eastern Germany this upper class is Saxon, and in Austria it is Swabian and Bavarian.

The introduction of Slavs in east Germany is known to be by infiltration and not by conquest. In the fourth century these Wends were called Venethi, Antes, and Sclaveni, and were described as strong in numbers but despised in war. Through the neglect of the Teutons they were allowed to range far and wide from their homes near the northeastern Carpathians, and to occupy the lands formerly belonging to the German nations, who had abandoned their country and flocked into the Roman Empire. Goth, Burgund, Lombard, and Vandal were replaced by the lowly Wend, and his descendants to-day form the privates in the east German regiments, while the officers are everywhere recruited from the Nordic upper class. The mediæval relation of these Slavic tribes to the dominant Teuton, is well expressed in the meaning—slave—which has been attached to their name in western languages.

The occupation of eastern Germany and Poland by the Slavs probably occurred from 400 A. D. to 700 A. D., but these Alpine elements were reinforced from the east and south from time to time during the succeeding centuries. Beginning early in the tenth century, under their Emperor, Henry the Fowler, the Saxons turned their attention eastward, and during the next two centuries they re-conquered and thoroughly Germanized all this section of Europe.

A similar series of changes in racial predominance took place in Russia where, in addition to a nobility largely Nordic, a section of the population is of ancient Nordic type, although the bulk of the peasantry consists of Alpine Slavs.

The Alpines in eastern Europe are represented by various branches of the Slavic nations. Their area of distribution was split into two sections by the occupation of the great Dacian plain by the Hungarians in about 900 A. D. These Magyars came from somewhere in eastern Russia beyond the sphere of Aryan speech, and their invasion separated the northern Slavs, known as Wends, Czechs, Slovaks, and Poles, from the southern Slavs, known as Serbs and Croats. These southern Slavs entered the Balkan Peninsula in the sixth century from the northeast, and to-day form the great mass of the population there.

The center of radiation of all these Slavic-speaking Alpines was located in the Carpathians, especially the Ruthenian districts of Galicia and eastward to the neighborhood of the Pripet swamps and the headwaters of the Dnieper in Polesia, where the Slavic dialects are believed to have developed, and whence they spread throughout Russia about the eighth century. These early Slavs were probably the Sarmatians of the Greek and Roman writers, and their name "Venethi" seems to have been a later designation. The original Proto-Slavic language, being Aryan, must have been at some distant date imposed by Nordics on the Alpines, but its development into the present Slavic tongues was chiefly the work of Alpines.

In other words, the expansion of the Alpines of the Slavic-speaking group seems to have occurred between 400 and 900 A. D., and they have spread in the East over areas which were originally Nordic, very much as the Teutons had previously overrun and submerged the earlier Alpines in the West. The Mongol, Tatar, and Turk, who invaded Europe much later, have little in common with the Alpine race, except the round skull. All these purely Asiatic types have been thoroughly absorbed and Europeanized, except in certain localities in Russia, especially in the east and south, where Mongoloid tribes have maintained their type either in isolated and relatively large groups, or side by side with their Slavic neighbors. In both cases the isolation is maintained by religious and social differences.

The Avars, also of Asiatic origin, preceded the Magyars in Hungary and the Slavs in the Balkans, but they have merged with the latter without leaving traces that can be identified, unless certain Mongoloid characters found in Bulgaria are of this origin.

The original physical type of the Magyars and the European Turks has now practically vanished, as a result of prolonged intermarriage with the original inhabitants of Hungary and the Balkans. These tribes have left little behind but their language, and in the case of the Turks, their religion. The brachycephalic Hungarians to-day resemble the Austrian-Germans much more than they do the Slavic-speaking populations surrounding them on the north and south, or the Rumanians on the east.

Following in the wake of the Avars, the Bulgarians appeared south of the Danube about the end of the seventh century, coming from eastern Russia, where the remnants of their kindred still persist along the Volga. To-day they conform physically in the western half of the country with the Alpine Serbs, and in the eastern half with the Mediterranean race, as do also the Rumanians of the Black Sea coast.

Little or nothing remains of the ancestral Bulgars except their name. Language, religion, and nearly, but not quite all, of the physical types have disappeared.

The early members of the Nordic race, in order to reach the Mediterranean world, had to pass through the Alpine populations, and must have absorbed a certain amount of Alpine blood. Therefore the Umbrians in Italy and the Gauls of western Europe, while predominantly Nordic, were more mixed with Alpine blood than were the Belgæ or Cymry, or their Teutonic successors, who, as Goths, Vandals, Burgundians, Helvetians, Alemanni, Saxons, Franks, Lombards, Danes, and Northmen, appear in history as pure Nordics of the Teutonic group.

In some portions of their range, notably Savoy and central France, the Alpine race is much less affected by Nordic influence than elsewhere, but on the other hand shows signs of a very ancient admixture with Mediterranean and even earlier elements. Brachycephalic Alpine populations in comparative purity still exist in the interior of Brittany, although almost completely surrounded by Nordic populations.

While the Alpines were everywhere swamped and driven to the fastnesses of the mountains, the warlike and restless nature of the Nordics has enabled the more stable Alpine population to slowly reassert itself, and Europe is probably much more Alpine to-day than it was fifteen hundred years ago.

The early Alpines made very large contributions to the civilization of the world, and were the medium through which many advances in culture were introduced from Asia into Europe. This race at the time of its first appearance in the west brought to the nomad hunters the knowledge of agriculture and of primitive pottery and of the domestication of animals, and thus made possible a great increase in population and the establishment of permanent settlements. Still later its final expansion was the means through which the knowledge of metals reached the Mediterranean and Nordic populations of the west and north. Upon the appearance on the scene of the Nordics the Alpine race lost its identity and sank to the subordinate and obscure position which it still occupies.

In western Asia members of this race are entitled to the honor of the earliest civilization of which we have knowledge, namely, that of Sumer and its northerly neighbor, Accad in Mesopotamia. It is also the race of Susa, Elam, and Media. In fact, the whole of Mesopotamian civilization belongs to this race with the exception of later Babylonia and Assyria, which were Arabic and Semitic, and of Persia and the empire of the Kassites, which were Nordic and Aryan.

In classic, mediæval, and modern times the Alpines have played an unimportant part in European culture, and in western Europe they have been so thoroughly Nordicized that they exist rather as an element in Nordic race development than as an independent type. There are, however, many indications in current history which point to a great development of civilization in the Slavic branches of this race, and the world must be prepared to face, as one of the results of the present war, a great industrial and cultural expansion in Russia, perhaps based on military power.

# V
# THE MEDITERRANEAN RACE

The Mediterranean subspecies, formerly called the Iberian, is a relatively small, light boned, long skulled race, of brunet color becoming even swarthy in certain portions of its range. Throughout Neolithic times and possibly still earlier, it seems to have occupied, just as it does today, all the shores of the Mediterranean, including the coast of Africa from Morocco on the west to Egypt on the east. The Mediterraneans are the western members of a subspecies of man which forms a substantial part of the population of Persia, Afghanistan, Baluchistan, and Hindustan, with perhaps a southward extension into Ceylon.

The Aryanized Afghan and Hindu of northern India speak languages derived from Old Sanskrit, and are distantly related to the Mediterranean race. Aside from a common dolichocephaly these peoples are entirely distinct from the Dravidians of south India whose speech is agglutinative and who show strong evidence of profound mixture with the ancient negrito substratum of southern Asia.

Everywhere throughout the Asiatic portion of its range the Mediterranean race overlies an even more ancient negroid race. These negroids still have representatives among the Pre-Dravidians of India, the Veddahs of Ceylon, the Sakai of the Malay Peninsula, and the natives of the Andaman Islands.

This Mediterranean subspecies at the close of the Paleolithic spread from the basin of the Inland Sea northward by way of Spain throughout western Europe, including the British Isles, and, before the final expansion of the Alpines, was widely distributed up to and touching the domain of the Nordic dolichocephs. It did not cross the Alps from the south, but spread around the mountains across the Rhine into western Germany.

In all this vast range from the British Isles to Hindustan, it is not to be supposed that there is identity of race. Certain portions, however, of the populations of the countries throughout this long stretch do show in their physique clear indications of descent from a Neolithic race of a common original type, which we may call Proto-Mediterranean.

Quite apart from inevitable admixture with late Nordic and early Paleolithic elements, the little brunet Englishman has had perhaps ten thousand years of independent evolution during which he has undergone selection due to the climatic and physical conditions of his northern habitat. The result is that he has specialized far away from the Proto-Mediterranean race which contributed this blood originally to Britain, probably while it was still a part of continental Europe.

On the other end of the range of the Mediterranean species, this race in India has been crossed with Dravidians and with Pre-Dravidian negroids. The Mediterraneans in India have also had imposed upon them other ethnic elements which came over through the Afghan passes from the northwest. The resultant racial mixture in India has had its own line of specialization. Residence in the fertile but unhealthy river bottoms, the direct rays of a tropic sun, and competition with the immemorial autochthones have unsparingly weeded generation after generation, until the existing Hindu has little in common with the ancestral Proto-Mediterranean.

It is to the Mediterranean race in the British Isles that the English, Scotch, and Americans owe whatever brunet characters they possess. In central Europe it underlies the Alpine race, and, in fact, wherever this race is in contact with either the Alpines or the Nordics, it appears to represent the more ancient stratum of the population.

So far as we know, this Mediterranean type never existed in Scandinavia, and all brunet elements found there are to be attributed to introductions in historic times. Nor did the Mediterranean race ever enter or cross the high Alps as did the Nordics, at a much later date, on their way to the Mediterranean basin from the Baltic coasts.

The Mediterranean race with its Asiatic extensions is bordered everywhere on the north of its enormous range from Spain to India by round skulls, but there does not seem to be as much evidence of mixture between these two subspecies of man as there is between the Alpines and the Nordics.

Along its southern boundary the Mediterranean race is in contact with either the long skull negroes of Ethiopia, or the ancient negrito population of southern Asia. In Africa this race has drifted southward over the Sahara and up the Nile valley, and has modified the blood of the negroes in both the Senegambian and equatorial regions.

Beyond these mixtures of blood, there is absolutely no relationship between the Mediterranean race and the negroes. The fact that the Mediterranean race is long skulled as well as the negro, does not indicate relationship as has been suggested. Overemphasis of the importance of the skull shape as a somatological character can easily be misleading, and other unit characters than skull proportions must also be carefully considered in all determinations of race.

Africa north of the Sahara, from a zoological point of view, is now, and has been since early Tertiary times, a part of Europe. This is true both of animals and of the races of man. The Berbers of north Africa to-day are racially identical with the Spaniards and south Italians and the ancient Egyptians and their modern descendants, the fellaheen, are merely clearly marked varieties of this Mediterranean race.

The Egyptians fade off toward the south into the so-called Hamitic people (to use an obsolete name), and the infusion of negro blood becomes increasingly great, until we finally reach the pure negro. On the east in Arabia we find an ancient and highly specialized subdivision of the Mediterranean race, which has from time out of mind crossed the Red Sea and infused its blood into the negroes of east Africa.

To-day the Mediterranean race forms in Europe a substantial part of the population of the British Isles, the great bulk of the population of the Iberian Peninsula, nearly one-third of the population of France, Liguria, Italy south of the Apennines, and all the Mediterranean coasts and islands, in some of which, like Sardinia, it exists in great purity. It forms the substratum of the population of Greece and of the eastern coasts of the Balkan Peninsula. Everywhere in the interior, except in eastern Bulgaria and Rumania, it has been replaced by the South Slavs and by the Albanians, the latter a mixture of the ancient Illyrians and the Slavs.

In the British Isles the Mediterranean race represents the Pre-Nordic population and exists in considerable numbers in Wales and in certain portions of England, notably in the Fen districts to the north of London. In Scotland it is nearly obliterated, leaving behind only its brunetness as an indication of its former prevalence, though it is now often associated there with tall stature.

This is the race that gave the world the great civilizations of Egypt, of Crete, of Phoenicia including Carthage, of Etruria and of Mycenaean Greece. It gave us, when mixed and invigorated with Nordic elements, the most splendid of all civilizations, that of ancient Hellas, and the most enduring of political organizations, the Roman State.

To what extent the Mediterranean race entered into the blood and civilization of Rome, it is now difficult to say, but the traditions of the Eternal City, its love of organization, of law and military efficiency, as well as the Roman ideals of family life, loyalty, and truth, point clearly to a Nordic rather than to a Mediterranean origin.

The struggles in early Rome between Latin and Etruscan, and the endless quarrels between patrician and plebeian, arose from the existence in Rome, side by side, of two distinct and clashing races, probably Nordic and Mediterranean respectively. The northern qualities of Rome are in sharp contrast to the Levantine traits of the classic Greeks, whose volatile and analytical spirit, lack of cohesion, political incapacity, and ready resort to treason, all point clearly to southern and eastern affinities.

While very ancient, present for probably ten thousand years in western and southern Europe, and even longer on the south shore of the Mediterranean, nevertheless this race cannot be called purely European. The route of its migration along the north coast of Africa, and up the west coast of Europe, can be traced everywhere by its beautifully polished stone weapons and tools. The Megalithic monuments also are found in association with this race, and mark its line of advance in western Europe, although they extend beyond the range of the Mediterraneans into the domain of the Scandinavian Nordics. These huge stone structures were chiefly sepulchral memorials and appear to have been based on an imitation of the Egyptian funeral monuments. They date back to the first knowledge of the manufacture and use of bronze tools by the Mediterranean race, and they occur in great numbers, vast size, and considerable variety along the north coast of Africa and up the Atlantic seaboard through Spain, Brittany, and England to Scandinavia.

It is admitted that the various groups of the Mediterranean race did not speak, in the first instance, any form of Aryan tongue. These Aryan languages we know were introduced into the Mediterranean world from the north. We have in the Basque tongue to-day a survival of one of the Pre-Aryan languages, which were spoken by the Mediterranean population of the Iberian Peninsula before the arrival of the Aryan-speaking Gauls of Nordic race.

The language of these invaders was Celtic, and replaced over most of the country the ancient speech of the natives, only in turn to be superseded, along with the Phoenician spoken in some of the southern coast towns, by the Latin of the conquering Roman, and Latin, mixed with some small elements of Gothic construction and Arabic vocabulary forms the basis of modern Portuguese, Castilian, and Catalan.

The native Mediterranean race of the Iberian Peninsula quickly absorbed the blood of these conquering Gauls, just as it later diluted beyond recognition the vigorous physical characters of the Teutonic Vandals, Suevi, and Visigoths. A certain amount of Nordic blood still persists to-day in northwestern Spain, especially in Galicia and along the Pyrenees, as well as generally among the upper classes. The Romans left no evidence of their domination except in their language and religion; while the earlier Phoenicians on the coasts, and the later swarms of Moors and Arabs all over the peninsula, but chiefly in the south, were closely related by race to the native Iberians.

That portion of the Mediterranean race which inhabits southern

France occupies the territory of ancient Languedoc and Provence, and it was these Provengals who developed and preserved during the Middle Ages the romantic civilization of the Albigensians, a survival of classic culture, which was drowned in blood by a crusade from the north in the thirteenth century.

In North Italy only the coast of Liguria is occupied by the Mediterranean race. In the valley of the Po the Mediterraneans were the predominant race during the early Neolithic, but with the introduction of bronze the Alpines appear, and round skulls to this day prevail north of the Apennines. About 1100 B. C. the Nordic Umbrians and Oscans swept over the Alps from the northeast, conquered northern Italy and introduced their Aryan speech, which gradually spread southward. The Umbrian state was afterward overwhelmed by the Etruscans, who were of Mediterranean race, and who, by 800 B. C. had extended their empire northward to the Alps. In the sixth century B. C. new swarms of Nordics, coming this time from Gaul and speaking Celtic dialects, seized the valley of the Po, and in 390 B. C. these Gauls, reinforced from the north and under the leadership of Brennus, stormed Rome and completely destroyed Etruscan power. From that time onward the valley of the Po became known as Cisalpine Gaul. Mixed with Nordic elements, chiefly Gothic and Lombard, this population persists to this day, and is the backbone of modern Italy.

A similar movement of these same Gauls or Galatians, as the Greek world called them, starting from northern Italy, occurred a century later when these Nordics suddenly appeared before Delphi in Greece in 279 B. C., and then swept over into Asia Minor and founded the state called Galatia, which endured until Christian times.

South Italy, until its conquest by Rome, was Magna Graecia, and the population to-day retains many Pelasgian Greek elements. It is among these Hellenic remnants that artists search for the handsomest types of the Mediterranean race. In Sicily also the race is purely Mediterranean in spite of the admixture of types coming from the neighboring coasts of Tunis. These intrusive elements, however, were all of kindred race. Traces of Alpine elements in these regions and on the adjoining African coast are very scarce, and are to be referred to the great and final wave of round skull invasion which introduced bronze into Europe.

In Greece the Mediterranean Pelasgians, who spoke a non-Aryan tongue, were swamped by the Nordic Achaeans, who entered from the northeast according to tradition prior to 1250 B. C., probably between 1400 and 1300 B. C. There were also probably still earlier waves of these same Nordic invaders as far back as 1700 B.C., which was a period of migration throughout the ancient world. These Achaeans were armed with iron weapons of the Hallstatt culture, with which they conquered the bronze using natives. The two races, as yet unmixed, stand out in clear contrast in the Homeric account of the siege of Troy, which is generally assigned to the date of n 94 to 1184 B. C.

The same invasion that brought the Achaeans into Greece brought a related Nordic people to the coast of Asia Minor, known as Phrygians. Of this race were the Trojan leaders.

Both the Trojans and the Greeks were commanded by huge blond princes, the heroes of Homer, while the bulk of the armies on both sides was composed of little brunet Pelasgians, imperfectly armed and remorselessly butchered by the leaders on either side. The only common soldiers mentioned by Homer as of the same race as the heroes, were the Myrmidons of Achilles.

About the time that the Achaeans and the Pelasgians began to amalgamate, new hordes of Nordic barbarians, collectively called Hellenes, entered from the northern mountains and destroyed this old Homeric-Mycenaean civilization. This Dorian invasion took place a little before 1100 B. C. and brought in the three main Nordic strains of Greece, the Dorian, the Æolian and the Ionian groups, which remain more or less distinct and separate throughout Greek history. It is more than probable that this invasion or swarming of Nordics into Greece was part of the same general racial upheaval that brought the Umbrians and Oscans into Italy.

Long years of intense and bitter conflict follow between the old population and the newcomers, and when the turmoil of this revolution settled down, classic Greece appears. What was left of the Achaeans retired to the northern Peloponnesus, and the survivors of the early Pelasgian population remained in Messenia serving as helots their Spartan masters. The Greek colonies in Asia Minor were founded by refugees fleeing from these Dorian invaders.

The Pelasgian strain seems to have persisted best in Attica and the Ionian states. The Dorian Spartans appear to have retained more of the character of the northern barbarians than the Ionian Greeks, but the splendid civilization of Hellas was due to a fusion of the two elements, the Achaean and Hellene of Nordic, and the Pelasgian of Mediterranean race.

The contrast between Dorian Sparta and Ionian Athens, between the military efficiency, thorough organization, and sacrifice of the citizen for the welfare of the state, which constituted the basis of the Lacedæmonian power, and the Attic brilliancy, instability, and extreme development of individualism, is strikingly like the contrast between Prussia with its Spartan-like culture and France with its Athenian versatility.

To this mixture of the two races in classic Greece the Mediterranean Pelasgians contributed their Mycenaean culture and the Nordic Achaeans and Hellenes contributed their Aryan language, fighting efficiency, and the European aspect of Greek life.

The first result of a crossing of two such contrasted subspecies as the Nordic and Mediterranean races, has repeatedly been a new outburst of culture. This occurs as soon as the older race has imparted to the conquerors its civilization, and before the victors have allowed their blood to be swamped by mixture. This process seems to have happened several times in Greece.

Later, in 339 B. C., when the original Nordic blood had been hopelessly diluted by mixture with the ancient Mediterranean elements, Hellas fell an easy prey to Macedon. The troops of Philip and Alexander were Nordic and represented the uncultured but unmixed ancestral type of the Achaeans and Hellenes. Their unimpaired fighting strength was irresistible as soon as it was organized into the Macedonian phalanx, whether directed against their degenerate brother Greeks, or against the Persians, whose original Nordic elements had also by this time practically disappeared. When in its turn the pure Macedonian blood was impaired by intermixture with Asiatics, they, too, vanished, and even the royal Macedonian dynasties in Asia and Egypt soon ceased to be Nordic or Greek except in language and customs.

It is interesting to note that the Greek states in which the Nordic element was most predominant outlived the other states. Athens fell before Sparta, and Thebes outlived them both. Macedon in classic times was considered quite the most barbarous state in Hellas, and was scarcely recognized as forming part of Greece, but it was through the military power of its armies and the genius of Alexander that the Levant and western Asia became Hellenized. Alexander, with his Nordic features, aquiline nose, gently curling yellow hair, and mixed eyes, the left blue and the right very black, typifies this Nordic conquest of the Near East.

It is not possible to-day to find in purity the physical traits of the ancient race in the Greek-speaking lands and islands, and it is chiefly among the pure Nordics of Anglo-Norman type that there occur those smooth and regular classic features, especially the brow and nose lines, that were the delight of the sculptors of Hellas.

So far as modern Europe is concerned culture came from the south and not from the east, and to this Mediterranean subspecies is due the foundation of our civilization. The ancient Mediterranean world was of this race; the long-sustained civilization of Egypt, which endured during thousands of years of almost uninterrupted sequence; the brilliant Minoan Empire of Crete, which flourished between 4000 and 1200 B. C., and was the ancestor of the Mycenaean cultures of Greece, Cyprus, Italy, and Sardinia; the mysterious empire of Etruria, the predecessor and teacher of Rome; the Hellenic states and colonies throughout the Mediterranean and Black Seas; the maritime and mercantile power of Phoenicia and its mighty colony, imperial Carthage; all were the creation of this race. The sea empire of Crete, when its royal palace at Cnossos was burned by the 'sea peoples' of the north, passed to Tyre, Sidon, and Carthage, and from them to the Greeks, so that the early development of the art of navigation is to be attributed to this race, and from them the north, centuries later, learned its maritime architecture.

Even though the Mediterranean race has no claim to the invention of the synthetic languages, and though it played a relatively small part in the development of the civilization of the Middle Ages or of modern times, nevertheless to it belongs the chief credit of the classic civilization of Europe, in the sciences, art, poetry, literature, and philosophy, as well as the major part of the civilization of Greece, and a very large share in the Empire of Rome.

In the Eastern Empire the Mediterraneans were the predominant factor under the guise of Byzantine Greeks. Owing to the fact that our histories have been written under the influence of Roman orthodoxy, and because in the eyes of the Frankish Crusaders the Byzantine Greeks were heretics, they have been regarded by us as degenerate cowards.

But throughout the Middle Ages Byzantium represented in unbroken sequence the Empire of Rome in the East, and as the capital of that empire it held Mohammedan Asia in check for nearly a thousand years. When at last in 1453 the imperial city, deserted by western Christendom, was stormed by the Ottoman Turks, and Constantine, last of Roman Emperors, fell sword in hand, there was enacted one of the greatest tragedies of all time.

With the fall of Constantinople the Empire of Rome passes finally from the scene of history, and the development of civilization is transferred from Mediterranean lands and Mediterranean race to the North Sea and the Nordic race.

# VI
# THE NORDIC RACE

We have shown that the Mediterranean race entered Europe from the south and forms part of a great group of peoples extending into southern Asia, that the Alpine race came from the east through Asia Minor and the valley of the Danube, and that its present European distribution is merely the westernmost point of an ethnic pyramid, the base of which rests solidly on the round skulled peoples of the great plateaux of central Asia. Both of these races are, therefore, western extensions of Asiatic subspecies, and neither of them can be considered as exclusively European.

With the remaining race, the Nordic, however, the case is different. This is a purely European type, and has developed its physical characters and its civilization within the confines of that continent. It is, therefore, the *Homo europœus*, the white man par excellence. It is everywhere characterized by certain unique specializations, namely, blondness, wavy hair, blue eyes, fair skin, high, narrow and straight nose, which are associated with great stature, and a long skull, as well as with abundant head and body hair.

This abundance of hair is an ancient and generalized character which the Nordics share with the Alpines of both Europe and Asia, but the light colored eyes and light colored hair are characters of relatively recent specialization and consequently highly unstable.

The pure Nordic race is at present clustered around the shores of the Baltic and North Seas, from which is has spread west and south and east in every direction, fading off gradually into the two preceding races.

The centre of its greatest purity is now in Sweden, and there is no doubt that at first the Scandinavian Peninsula, and later the immediately adjoining shores of the Baltic, were the centres of radiation of the Teutonic or Scandinavian branch of this race.

The population of Scandinavia has been composed of this Nordic subspecies from the beginning of Neolithic times, and Sweden to-day represents one of the few countries which has never been overwhelmed by foreign conquest, and in which there has been but a single racial type from the beginning. This nation is unique for its unity of race, language, religion, and social ideals.

Southern Scandinavia only became fit for human habitation on the retreat of the glaciers about twelve thousand years ago and apparently was immediately occupied by the Nordic race. This is one of the few geological dates which is absolute and not relative. It rests on a most interesting series of computations made by Baron DeGeer, based on an actual count of the laminated deposits of clay laid down annually by the retreating glaciers, each layer representing the summer deposit of the subglacial stream.

The Nordics first appear at the close of the Paleolithic along the coasts of the Baltic. The earliest industry discovered in this region is known as the Maglemose, found in Denmark and elsewhere around the Baltic, and is probably the culture of the Proto-Teutonic branch of the Nordic race. No human remains have as yet been found.

The vigor and power of the Nordic race as a whole is such that it could not have been evolved in so restricted an area as southern Sweden, although its Teutonic section did develop there in comparative isolation. The Nordics must have had a larger field for their specialization, and a longer period for their evolution, than is afforded by the limited time which has elapsed since Sweden became habitable. For the development of so marked a type there is required a continental area isolated and protected for long ages from the intrusion of other races. The climatic conditions must have been such as to impose a rigid elimination of defectives through the agency of hard winters and the necessity of industry and foresight in providing the year's food, clothing, and shelter during the short summer. Such demands on energy, if long continued, would produce a strong, virile, and self-contained race which would inevitably overwhelm in battle nations whose weaker elements had not been purged by the conditions of an equally severe environment.

An area conforming to these requirements is offered by the forests and plains of eastern Germany, Poland, and Russia. It was here that the Proto-Nordic type evolved, and here their remnants are found. They were protected from Asia on the east by the then almost continuous water connections across eastern Russia between the White Sea and the old Caspian-Aral Sea.

During the last glacial advance (the Würm glaciation), which, like the preceding glacial advances, is believed to have been a period of land depression, the White Sea extended far to the south of its present limits, while the enlarged Caspian Sea, then and long afterward connected with the Sea of Aral, extended northward to the great bend of the Volga. The intermediate area was studded with large lakes and morasses. Thus an almost complete water barrier of shallow sea, located just west of the low Ural Mountains, separated Europe from Asia during the Wiirm glaciation and long afterward. The broken connection was restored just before the dawn of history by the slight elevation of the land and the shrinking of the Caspian-Aral Sea through increasing desiccation which left its present surface below sea level.

An important element in the isolation of this Nordic cradle on the south is the fact that from the earliest times down to this day the pressure of population has everywhere been from the bleak and sterile north southward and eastward into the sunny and enervating lands of France, Italy, Greece, Persia, and India.

In these forests and steppes of the north, the Nordic race gradually evolved in isolation, and at a very early date occupied the Scandinavian Peninsula, together with much of the land now submerged under the Baltic and North Seas.

Nordic strains form everywhere a substratum of population throughout Russia and underlie the round skull Slavs who first appear a little over a thousand years ago as coming, not from the direction of Asia, but from south Poland. Burial mounds called kurgans are widely scattered throughout Russia from the Carpathians to the Urals, and contain numerous remains of a dolichocephalic race; in fact, more than three-fourths of the skulls are of this type. Round skulls first become numerous in ancient Russian graveyards about 900 A. D., and soon increase to such an extent that in the Slavic period from the ninth to the thirteenth centuries one-half of the skulls were brachycephalic, while in modern cemeteries the proportion of round skulls is still greater. This ancient Nordic element, however, still forms a very considerable portion of the population of northern Russia and contributes the blondness and the red-headedness so characteristic of the Russian of to-day. As we leave the Baltic coasts the Nordic characters fade out both toward the south and east. The blond element in the nobility of Russia is of later Scandinavian and Teutonic origin.

When the seas which separated Russia from Asia dried up, and when the isolation and exacting climate of the north had done their work and produced the vigorous Nordic type, these men burst upon the southern races, conquering east, south, and west. They brought with them from the north the hardihood and vigor acquired under the rigorous selection of a long winter season, and vanquished in battle the inhabitants of older and feebler civilizations, only in their turn to succumb to the softening influences of a life of ease and plenty in their new homes.

The earliest appearance in history of Aryan-speaking Nordics is our first dim vision of the Sacæ introducing the Sanskrit into India, the Cimmerians pouring through the passes of the Caucasus from the grass-lands of south Russia to invade the Empire of the Medes, and the Achaeans and Phrygians conquering Greece and the Ægean coast of

Asia Minor. About 1100 B. C. Nordics enter Italy as Umbrians and Oscans, and soon after cross the Rhine into Gaul. This western vanguard was composed of Celtic-speaking tribes which had long occupied those districts in Germany which lay south and west of the Teutonic-speaking Nordics, who at this early date were probably confined to Scandinavia and the immediate shores of the Baltic, and were beginning to press southward.

This first wave of Nordics seems to have swept westward along the sandy plains of northern Europe, entering France through the Low Countries. From this point as Goidels they spread north into Britain, reaching there about 800 B. C. As Gauls they conquered all France and pushed on south and west into Spain, and over the Maritime Alps into northern Italy, where they encountered their kindred Nordic Umbrians, who at an earlier date had crossed the Alps from the northeast. Other Celtic-speaking Nordics apparently migrated up the Rhine and down the Danube, and by the time the Romans came on the scene the Alpines of central Europe had been thoroughly Celticized. These tribes pushed eastward into southern Russia and reached the Crimea as early as the fourth century B.C. Mixed with the natives, they were called by the Greeks the Celto-Scyths. This swarming out of Germany of the first Nordics was during the closing phases of the Bronze Period, and was contemporary with, and probably caused by, the first great expansion of the Teutons from Scandinavia by way both of Denmark and the Baltic coasts.

These invaders were succeeded by a second wave of Celtic-speaking peoples, the Cymry, who drove their Goidelic predecessors still farther west and exterminated and absorbed them over large areas. These Cymric invasions occurred about 300-100 B. C., and were probably the result of the growing development of the Teutons and their final expulsion of the Celtic-speaking tribes from Germany. These Cymry occupied northern France under the name of Belgæ and invaded England as Brythons, and their conquests in both Gaul and Britain were only checked by the legions of Caesar.

These migrations are exceedingly hard to trace because of the confusion caused by the fact that Celtic speech is now found on the lips of populations in nowise related to the Nordics who first introduced it. But one fact stands out clearly, all the original Celtic-speaking tribes were purely Nordic.

What were the special physical characters of these tribes, in which they differed from their Teutonic successors, is now impossible to say, beyond the possible suggestion that in the British Isles the Scottish and Irish populations in which red hair and gray or green eyes are abundant have rather more of this Celtic strain in them than have the flaxen haired Teutons, whose china blue eyes are clearly not Celtic.

When the peoples called Gauls or Celts by the Romans, and Galatians by the Greeks, first appear in history, they are described in exactly the same terms as were later the Teutons. They were all gigantic barbarians with fair and very often red hair, then more frequent than to-day, with gray or fiercely blue eyes, and were thus clearly members of the Nordic subspecies.

The first Celtic-speaking nations with whom the Romans came in contact were Gaulish, and had probably incorporated much Alpine blood by the time they crossed the mountains into the domain of classic history. The Nordic element had become still weaker by absorption from the conquered populations, when at a later date the Romans broke through the ring of Celtic nations and came into contact with the purely Nordic Cymry and Teutons.

After these early expansions of Gauls and Cymry, the Teutons appear upon the scene. Of the pure Teutons within the ken of history, it is not necessary to mention more than the most important of the long series of conquering tribes.

The greatest of them all were perhaps the Goths, who came originally from the south of Sweden and were long located on the opposite German coast, at the mouth of the Vistula. From here they crossed Poland to the Crimea, where they were known in' the first century. Three hundred years later they were driven westward by the Huns and forced into the Dacian plain and over the Danube into the Roman Empire. Here they split up; the Ostrogoths after a period of subjection to the Huns on the Danube, ravaged the European provinces of the Eastern Empire, conquered Italy, and founded there a great but short-lived nation. The Visigoths occupied much of Gaul and then entered Spain, driving the Vandals before them into Africa. The Teutons and Cimbri destroyed by Marius in southern Gaul about too B. C.; the Gepidae; the Alans; the Suevi; the Vandals; the Helvetians; the Alemanni of the upper Rhine; the Marcomanni; the Saxons; the Batavians; the Frisians; the Angles; the Jutes, the Lombards and the Heruli of Italy; the Burgundians of the east of France; the Franks of the lower Rhine; the Danes; and latest of all, the Norse Vikings, swept through history. Less well known but of great importance, are the Varangians, who, coming from Sweden in the ninth and tenth centuries, conquered the coast of the Gulf of Finland and much of White Russia, and left there a dynasty and aristocracy of Norse blood. In the tenth and eleventh centuries they were the rulers of Russia.

The traditions of Goths, Vandals, Lombards, and Burgundians all point to Sweden as their earliest homeland, and probably all the pure Germanic tribes came originally from Scandinavia and were closely related.

When these Teutonic tribes poured down from the Baltic coasts, their Celtic-speaking Nordic predecessors were already much mixed with the underlying populations, Mediterranean in the west and Alpine in the south. These "Celts" were not recognized by the Teutons as kin in any sense, and were all called Welsh or foreigners. From this word are derived the names "Wales," "Cornwales" or "Cornwall," "Valais," "Walloons," and "Wallachian" or "Vlach."

# VII
# TEUTONIC EUROPE

No proper understanding is possible of the meaning of the history of Christendom, or full appreciation of the place in it of the Teutonic Nordics, without a brief review of the events in Europe of the last two thousand years.

When Rome fell and changed trade conditions necessitated the transfer of power from its historic capital in Italy to a strategic situation on the Bosporus, western Europe was definitely and finally abandoned to its Germanic invaders. These same barbarians swept up again and again to the Propontis, only to recoil before the organized strength of the Byzantine Empire, and the walls of Mikklegard.

Until the coming of the Alpine Slavs the Eastern Empire still held in Europe the Balkan Peninsula and much of the eastern Mediterranean. The Western Empire, however, collapsed utterly under the impact of hordes of Nordic Teutons at a much earlier date. In the fourth and fifth centuries of our era, north Africa, once the empire of Carthage, had become the seat of the kingdom of Teutonic Vandals. Spain fell under the control of the Visigoths, and Lusitania, now Portugal, under that of the Suevi. Gaul was Visigothic in the south and Burgundian in the east, while the Frankish kingdom dominated the north until it finally absorbed and incorporated all the territories of ancient Gaul and made it the land of the Franks.

Italy fell under the control first of the Ostrogoths and then of the Lombards. The purely Teutonic Saxons, with kindred tribes, conquered the British Isles, and meanwhile the Norse and Danish Scandinavians contributed large elements to all the coast populations as far south as Spain, and the Swedes organized in the eastern Baltic what is now Russia.

Thus when Rome passed, all Europe had become superficially Teutonic. At first these Teutons were isolated and independent tribes, bearing some shadowy relation to the one organized state they knew, the Empire of Rome. Then came the Mohammedan invasion, which reached western Europe from Africa and destroyed the Visigothic kingdom. The Moslems swept on unchecked until their light horsemen dashed themselves to pieces against the heavy armed cavalry of Charles Martel and his Franks at Tours in 732 A. D.

The destruction of the Vandal kingdom by the armies of the Byzantine Empire; the conquest of Spain by the Moors, and finally the overthrow of the Lombards by the Franks were all greatly facilitated by the fact that these barbarians, Vandals, Goths, Suevi, and Lombards, with the sole exception of the Franks, were originally Christians of the Arian or Unitarian confession, and as such were regarded as heretics by their Orthodox Christian subjects. The Franks alone were converted from heathenism directly into the Trinitarian faith to which the old populations of the Roman Empire adhered. From this orthodoxy of the Franks arose the close relation between France, "the eldest daughter of the church," and the papacy, a connection which lasted for more than a thousand years—in fact nearly to our own day.

With the Goths eliminated, western Christendom became Frankish. In the year 800 A. D. Charlemagne was crowned at Rome and re-established the Roman Empire in the west, which included all Christendom outside of the Byzantine Empire. In some form or shape this Roman Empire endured until the beginning of the nineteenth century, and during all that time it formed the basis of the political concept of European man.

This same concept lies to-day at the root of the imperial idea. The Kaiser, Tsar, and Emperor all take their name, and in some way trace their title, from Caesar and the Empire. Charlemagne and his successors claimed, and often exercised, over-lordship as to all the other continental Christian nations, and when the Crusades began it was the German Emperor who led the Frankish hosts against the Saracens. Charlemagne was a German Emperor, his capital was at Aachen, within the present limits of the German Empire, and the language of his court was German. For several centuries after the conquest of Gaul by the Franks, their Teutonic tongue held its own against the Latin speech of the Romanized Gauls.

The history of all Christian Europe is in some degree interwoven with this Holy Roman Empire. Though the Empire was neither holy nor Roman, but altogether secular and Teutonic, it was, nevertheless, the central core of Europe for ages. Holland and Flanders, Lorraine and Alsace, Burgundy and Luxemberg, Lombardy and Venezia, Switzerland and Austria, Bohemia and Styria are states which were originally component parts of the Empire, although many of them have since been torn away by rival nations or have become independent, while much of northern Italy remained under the sway of Austria within the memory of living men.

The Empire wasted its strength in imperial ambitions and foreign conquests instead of consolidating, organizing, and unifying its own territories, and the fact that the imperial crown was elective for many generations before it became hereditary in the House of Hapsburg, checked the unification of Germany during the Middle Ages.

A strong hereditary monarchy such as those which arose in England and in France would have anticipated the Germany of to-day by a thousand years and made it the predominant state in Christendom, but disruptive elements, in the persons of great territorial dukes, were successful throughout its history in preventing an effective concentration of power in the hands of the Emperor.

That the German Emperor was regarded, though vaguely, as the overlord of all Christian monarchs was clearly indicated when Henry VIII of England and Francis I of France appeared as candidates for the imperial crown against Charles of Spain, afterward the Emperor Charles V.

Europe was Germany, and Germany was Europe, predominantly, until the Thirty Years' War. This war was perhaps the greatest catastrophe of all the ghastly crimes committed in the name of religion. It destroyed an entire generation, taking each year for thirty years the finest manhood of the nations.

Two-thirds of the population of Germany was destroyed, in some states such as Bohemia three-fourths of the inhabitants were killed or exiled, while out of 500,000 inhabitants in Würtemberg there were only 48,000 left at the end of the war. Terrible as this loss was, the destruction did not fall equally on the various races and classes in the community. It bore, of course, most heavily upon the big blond fighting man, and at the end of the war the German states contained a greatly lessened proportion of Nordic blood. In fact from that time on the purely Teutonic race in Germany has been largely replaced by the Alpine types in the south, and by the Wendish and the Polish types in the east. This change of race in Germany has gone so far that it has been computed that out of the 70,000,000 inhabitants of the German Empire, only 9,000,000 are purely Teutonic in coloration, stature, and skull characters. The rarity of pure Teutonic and Nordic types among the German immigrants to

America in contrast to its almost universal prevalence among those from Scandinavia is traceable to the same cause.

In addition, the Thirty Years' War virtually destroyed the land owning yeomanry and lesser gentry formerly found in mediæval Germany as numerously as in France or in England. The religious wars of France, while not as devastating to the nation as a whole as was the Thirty Years' War in Germany, nevertheless greatly weakened the French cavalier type, the "petite noblesse de province." In Germany this class had flourished, and throughout the Middle Ages contributed great numbers of knights, poets, thinkers, great artists and artisans who gave charm and variety to European society. But as said, this section of the population was practically exterminated in the Thirty Years' War, and the class of gentlemen practically vanishes from German history from that time on.

When the Thirty Years' War was over there remained in Germany nothing except the brutalized peasantry, largely of Alpine derivation in the south and east, and the high nobility which turned from the toils of endless warfare to mimic on a small scale the court of Versailles. It has taken Germany two centuries to recover her vigor, her wealth, and her aspirations to a place in the sun.

During these years Germany was a political nonentity, a mere congery of petty states bickering and fighting with each other, claiming and owning only the Empire of the Air as Napoleon happily phrased it, and meantime France and England founded their colonial empires beyond the seas.

When, in the last generation, Germany became unified and organized, she found herself not only too late to share in these colonial enterprises, but also lacking in much of the racial element, and still more lacking in the very classes which were her greatest strength and glory before the Thirty Years' War. To-day the ghastly rarity in the German armies of chivalry and generosity toward women, and of knightly protection and courtesy toward the prisoners or wounded, can be largely attributed to this annihilation of the gentle classes. The Germans of today, whether they live on the farms or in the cities, are for the most part, descendants of the peasants who survived, not of the brilliant knights and sturdy foot soldiers who fell in that mighty conflict. Knowledge of this great past when Europe was Teutonic, and memories of the shadowy grandeur of the Hohenstaufen Emperors, who, generation after generation, led Teutonic armies over the Alps to assert their title to Italian provinces, have played no small part in modern German consciousness.

These traditions and the knowledge that their own religious dissensions swept them from the leadership of the European world, lie at the base of the German imperial ideal of to-day, and it is for this ideal that the German armies are dying, just as did their ancestors for a thousand years under their Fredericks, Henrys, Conrads, and Ottos.

But the Empire of Rome and the Empire of Charlemagne are no more, and the Teutonic type is divided almost equally between the contending forces in this world war. Germany is too late, and is limited to a destiny fixed and ordained for her on the fatal day in 1618 when the Hapsburg Ferdinand forced the Protestants of Bohemia into revolt.

Although as a result of the Thirty Years' War the German Empire' is far less Nordic than in the Middle Ages, the north of Germany is still Teutonic throughout, and in the east and south the Alpines have been thoroughly Germanized with an aristocracy and upper class of pure Teutonic blood.

# VIII
# THE EXPANSION OF THE NORDICS

The men of Nordic blood to-day form all the population of Scandinavian countries, as also a majority of the population of the British Isles, and are almost pure in type in Scotland and eastern and northern England. The Nordic realm includes all the northern third of France, with extensions into the fertile southwest; all the rich lowlands of Flanders; all Holland; the northern half of Germany, with extensions up the Rhine and down the Danube; and the north of Poland, and of Russia. Recent calculations show that there are about 90,000,000 of purely Nordic physical type in Europe out of a total population of 420,000,000.

Throughout southern Europe a Nordic nobility of Teutonic type everywhere forms the old aristocratic and military classes, or what now remains of them. These aristocrats, by as much as their blood is pure, are taller and blonder than the native populations, whether these be Alpine in central Europe or Mediterranean in Spain or in the south of France and Italy.

The countries speaking Low German dialects are almost purely Nordic, but the populations of High German speech are very largely Teutonized Alpines, and occupy lands once Celtic-speaking. The main distinction between the two dialects is the presence of a large number of Celtic elements in High German.

In northern Italy there is a large amount of Nordic blood. In Lombardy, Venice, and elsewhere throughout the country the aristocracy is blonder and taller than the peasantry, but the Nordic element in Italy has declined noticeably since the Middle Ages. From Roman times onward for a thousand years the Teutons swarmed into northern Italy, through the Alps, chiefly by way of the Brenner Pass. With the stoppage of these Nordic invasions this strain seems to have grown less all through Italy.

In the Balkan Peninsula there is little to show for the floods of Nordic blood that have poured in for the last 3,500 years, beginning with the Achaeans of Homer, who first appeared en masse about 1400 B. C., and were followed successively by the Dorians, Cimmerians, and Gauls, down to the Goths and the Varangians of Byzantine times.

The tall stature of the population along the Illyrian Alps from the Tyrol to Albania on the south, is undoubtedly of Nordic origin, and dates from some of these early invasions, but these Illyrians have been so crossed with Slavs that all other blond elements have been lost, and the existing population is essentially of brachycephalic Alpine type. What few remnants of blondness occur in this district, more particularly in Albania, are probably to be attributed to later infiltrations, as are the so-called Frankish elements in Bosnia. In Russia and in Poland the Nordic stature, blondness, and long skull grow less and less pronounced as one proceeds south and east from the Gulf of Finland.

It would appear that in all those parts of Europe outside of its natural habitat, the Nordic blood is on the wane from England to Italy, and that the ancient, acclimated, and primitive populations of Alpine and Mediterranean race are subtly reasserting their long lost political power through a high breeding rate and democratic institutions.

In western Europe the first wave of the Nordic tribes appeared about three thousand years ago, and was followed by other invasions with the Nordic element becoming stronger until after the fall of Rome whole tribes moved into its provinces Germanizing them more or less for varying lengths of time.

These incoming Nordics intermarried with the native populations and were gradually bred out, and the resurgence of the old native stock has proceeded steadily since the Frankish Charlemagne destroyed the Lombard kingdom, and is proceeding with unabated vigor to-day. This process has been greatly accelerated in western Europe by the crusades and the religious and Napoleonic wars. The world war, now in full swing with its toll of millions, will leave Europe much poorer in Nordic blood. One of its most certain results will be the partial destruction of the aristocratic classes everywhere in northern Europe. In England the nobility has already suffered in battle more than in any century since the Wars of the Roses. This will tend to realize the standardization of type so dear to democratic ideals. If equality cannot be obtained by lengthening and uplifting the stunted of body and of mind, it can be at least realized by the destruction of the exalted of stature and of soul. The bed of Procrustes operates with the same fatal exactness when it shortens the long as when it stretches the undersized.

The first Nordics in Spain were the Gauls who crossed the Pyrenees about the seventh century before our era, and introduced Aryan speech into the Iberian Peninsula. They quickly mixed with Mediterranean natives and the composite Spaniards were called Celtiberians by the Romans.

In Portugal and Spain there are in the physical structure of the population few traces of these early Celtic-speaking Nordic invaders, but the Suevi, who a thousand years later occupied parts of Portugal, and the Vandals and Visigoths who conquered and held Spain for 300 years, have left some small evidence of their blood, and in the provinces of northwestern Spain a considerable percentage of light colored eyes reveals these Nordic elements in the population.

Deep seated Castilian traditions associate aristocracy with blondness, and the *sangre azul*, or blue blood of Spain, refers to the blue eye of the Goth, whose traditional claim to lordship is also shown in the Spanish name for gentleman, "hidalgo," or son of the Goth.

As long as this Gothic nobility controlled the Spanish states during the endless crusades against the Moors, Spain belonged with the Nordic kingdoms, but when their blood became impaired by losses in wars waged outside of Spain and in the conquest of the Americas, the sceptre fell from this noble race into the hands of the little, dark Iberian, who had not the physical vigor or the intellectual strength to maintain the world empire built up by the stronger race.

The splendid conquistadores of the New World were of Nordic type, but their pure stock did not long survive their new surroundings, and to-day they have vanished utterly, leaving behind them only their language and their religion. After considering well these facts we shall not have to search further for the causes of the collapse of Spain.

Gaul at the time of Cæsar's conquest was under the rule of the Nordic race, which furnished the bulk of the population of the north as well as the military classes elsewhere, and the power and vigor of the French nation have been based on this blood and its later reinforcements. In fact, in the Europe of to-day the amount of Nordic blood in each nation is a very fair measure of its strength in war and standing in civilization.

When, about 1000 B.C., the first Nordics crossed the lower Rhine they found the Mediterranean race in France everywhere overwhelmed by an Alpine population, except in the south, and before the time of Caesar the Celtic language of these invaders, which was related to the Goidelic language still spoken in parts of Ireland and in the Scotch Highlands, had been imposed upon the entire population, and the whole country had been saturated with Nordic blood. These earliest Nordics in the west were known to the ancient world as Gauls. These Gauls or "Celts," as they were called by Caesar, occupied in his day the centre of France. The actual racial complexion of this part of France was overwhelmingly Alpine then and is so now, but this population was Celticized thoroughly by the Gauls, just as it was Latinized as completely at a later date by the Romans.

The northern third of France, that is, above Paris, was inhabited in Caesar's time by the Belgæ, a Nordic people of the Cymric division of Celtic speech. They were largely of Teutonic blood, and in fact should be regarded as the immediate forerunners of the Germans, and they probably represent the early Teutons who had crossed from Sweden and adopted the Celtic speech of their Nordic kindred whom they found on the mainland. These Belgæ had followed the earlier Goidels across Germany into Britain and Gaul, and were rapidly displacing their Nordic predecessors, who by this time were much weakened by mixture with the autochthones, when Rome appeared upon the scene and set a limit to their conquests by the Pax Romana.

The Belgæ of the north of France and the Low Countries were the bravest of the peoples of Gaul, according to Caesar's well-known remark, but the claim of the Belgians of to-day to descent from this race is without basis and rests solely on the fact that the present Kingdom of Belgium, which only became independent and assumed its proud name in 1830, occupies a small and relatively unimportant corner of the land of the Belgæ. The Flemings of Belgium are Nordic Franks speaking a Low German tongue, and the Walloons are Alpines whose language is an archaic French.

The Belgæ and the Goidelic remnants of Nordic blood in the centre of Gaul, taken together constituted probably only a minority in blood of the population, but were everywhere the military and ruling classes. These Nordic elements were later reinforced by powerful Teutonic tribes,- namely, Vandals, Visigoths, Alans, Saxons, Burgundians, and most important of all, the Franks of the lower Rhine, who founded modern France and made it for long centuries the "*grand nation*" of Christendom.

The Frankish dynasties long after Charlemagne were of purely Teutonic blood, and the aristocratic land owning and military classes down to the great Revolution were everywhere of this type, which by the time of the creation of the Frankish kingdom had incorporated all the other Nordic elements of old Roman Gaul, both Gaulish and Belgic.

The last invasion of Teutonic-speaking barbarians was that of the Danish Northmen, who were, of course, of pure Nordic blood, and who conquered and settled Normandy in 911 A. D. No sooner had the barbarian invasions ceased than the ancient aboriginal blood strains, Mediterranean and Alpine, and elements derived from Paleolithic times, began a slow and steady recovery. Step by step, with the reappearance of these primitive and deep rooted stocks, the Nordic element in France declined, and with it the vigor of the nation.

The chief historic events of the last thousand years have hastened this process, and the fact that the Nordic element everywhere forms the fighting section of the community caused the loss in war to fall disproportionately as among the three races in France. The religious wars greatly weakened the Nordic provincial nobility, which was at first largely Protestant, and the process of exterminating the upper classes was completed by the Revolutionary and Napoleonic wars. These last wars are said to have shortened the stature of the French by four inches; in other words, the tall Nordic strain was killed off in greater proportions than the little brunet.

When by universal suffrage the transfer of power was completed from a Nordic aristocracy to lower classes predominantly of Alpine and Mediterranean extraction, the decline of France in international power set in.

The survivors of the aristocracy, being stripped of political power and to a large extent of wealth, quickly lost their caste pride and committed class suicide by mixing their blood with inferior breeds. One of the most conspicuous features of many of the French nobility of to-day is the strength of the Levantine and Mediterranean strain in them. Being, for political reasons, ardently clerical, the nobility welcomes recruits of any racial origin, as long as they bring with them money and devotion to the Church.

The loss in war of the best breeding stock through death, wounds, or absence from home has been clearly shown in France. The conscripts who were examined for military duty in 1890-2 were those descended in a large measure from the military rejects and other stay-at-homes during the Franco-Prussian War. In Dordogne this contingent showed seven per cent more deficient statures than the normal rate. In some cantons this unfortunate generation was in height an inch below the recruits of preceding years, and in it the exemptions for defective physique rose from the normal six per cent to sixteen per cent.

When each generation is decimated or destroyed in turn, a race can be injured beyond recovery, but it more frequently happens that the result is the annihilation of an entire class, as in the case of the German gentry in the Thirty Years' War. Desolation of wide districts often resulted from the plagues and famines which followed the armies in old days, but deaths from these causes fall most heavily on the weaker part of the population. The loss of valuable breeding stock is far more serious when wars are fought with volunteer armies of picked men than with conscript armies, because in the latter cases the loss is more evenly spread over the whole nation. Before England resorted in the present war to universal conscription the injury to her more desirable and patriotic classes was much more pronounced than in Germany, where all types and ranks are called to arms.

In the British Isles we find, before the arrival of the Nordic race, a Mediterranean population and no perceptible element of Alpine blood, so that we have to deal with only two of the main races instead of all three as in France. In Britain there are, as elsewhere, representatives of earlier races, but the preponderant strain of blood was Mediterranean before the first arrival of the Aryan-speaking Nordics.

Ireland was connected with Britain and Britain with the continent until times very recent in a geological sense. The depression of the Channel coasts is progressing rapidly to-day, and is known to have been substantial during historic times. The close parallel in blood and culture between England and the opposite coasts of France also indicates a very recent land connection, probably in Neolithic times. Men either walked from the continent to England and from England to Ireland, or they paddled across in primitive boats or coracles. The art of ship-building, or even archaic navigation, cannot go much further back than late Neolithic times.

The tribes of Celtic speech came to the British Isles in two distinct waves. The earlier invasion of the Goidels arrived in England with a culture of bronze about 800 B. C., and in Ireland two centuries later, and was part of the same movement which brought the Gauls into France. The later conquest was by the Cymric-speaking Belgæ who were equipped with iron weapons. It began in the third century B. C., and was still going on in Caesar's time. These Cymric Brythons found the early Goidels, with the exception of the aristocracy, much weakened by intermixture with the Mediterranean natives, and would probably have destroyed all trace of Goidelic speech in Ireland and Scotland, as they actually did in England, if the Romans had not intervened. The Brythons reached Ireland in small numbers only in the second century B. C.

These Nordic elements in Britain, both Goidelic and Brythonic, were in a minority during Roman times, and the ethnic complexion of the island was not much affected by the Roman occupation, as the legions stationed there represented the varied racial stocks of the Empire.

After the Romans abandoned Britain, and about 400 A. D., floods of pure Nordics poured into the islands for nearly six centuries, arriving in the north as the Norse pirates, who made Scotland Scandinavian, and in the east as Teutonic Saxons and Angles, who founded England.

The Angles came from somewhere in central Jutland, and the Saxons came from coast lands immediately at the base of the Danish Peninsula. All these districts were then, and are now, purely Teutonic; in fact, this is part of old Saxony, and is to-day the core of Germany.

These Saxon districts sent out at that time swarms of invaders not only into England but into France and over the Alps into Italy, just as at a much later period the same land sent swarming colonies into Hungary and Russia.

The same Saxon invaders passed down the Channel coasts, and traces of their settlement on the mainland remain to this day in the Cotentin district around Cherbourg. Scandinavian sea peoples, called Danes or Northmen, swarmed over as late as 900 A. D. and conquered all eastern England. This Danish invasion of England was the same that brought the Northmen, or Normans, into France. In fact the occupation of Normandy was probably by Danes, and the conquest of England was largely the work of Norsemen, as Norway at that time was under Danish kings.

Both of these invasions, especially the later one, swept around the greater island and inundated Ireland, driving the aborigines and their Celtic-speaking masters into the bogs and islands of the extreme west.

The blond Nordic element to-day predominates in Ireland as much as in England. It is derived, to some extent, from the early invaders of Celtic speech, but the Goidelic element has been in Ireland, as in England and Scotland, very largely absorbed by the Iberian substratum of the population, and is found to-day rather in the form of Nordic characters in brunets, than as the pure blond individuals who represent later and purer Nordic strains. The combination of black Iberian hair with blue or gray Nordic eyes is frequently found in Ireland and also in Spain, and in both these countries is greatly admired for its beauty.

The tall, blond Irishmen are to-day chiefly Danish with the addition of English, Norman, and Scotch elements, which have poured into the lesser island for a thousand years, and have imposed the English speech upon it. The more primitive and ancient elements in Ireland have always showed great ability to absorb newcomers, and during the Middle Ages it was notorious that the Norman and English colonists quickly sank to the cultural level of the natives. Indications of Paleolithic man appear in Ireland frequently as unit characters, as well as individuals. Being, like Brittany, situated on the extreme western outposts of Eurasia, it has more than its share of generalized and low types surviving in the living populations, and these types, the Firbolgs, have imparted a distinct and very undesirable aspect to a large portion of the inhabitants of the west and south, and have greatly lowered the intellectual status of the population as a whole.

In England much the same ethnic elements are present, namely the Nordic and the Mediterranean. There is, especially in Wales and in the west central counties of England, a large substratum of ancient Mediterranean blood, but the later coming Nordic elements are everywhere imposed upon it.

Scotland is by race Anglian in the south and Norse in the Highlands, with underlying Goidelic and Brythonic elements which are exceedingly hard to identify.

The Nordic species of man in his various races, but chiefly Teutonic, made Gaul the land of the Franks, and made Britain the land of the Angles, and the Englishmen who built the British Empire and founded America were of the Nordic and not of the Mediterranean type.

One of the most vigorous Nordic elements in France, England, and America was contributed by the Normans, and its influence on the development of these countries cannot be ignored. The descendants of the Danish and Norse Vikings who settled in Normandy as Teutonic-speaking heathen, and who as Normans crossed over to Saxon England and conquered it in 1066, are among the finest and noblest examples of the Nordic race. Their only rivals in these characters were the early Goths.

This Norman strain, while purely Nordic, seems to have been radically different in its mental makeup, and to some extent in its physical detail, from the Saxons of England, and also from the kindred Scandinavians on the continent.

The Normans seem to have been "*fine race*," to use a French idiom, and are often characterized by a tall, slender figure, proud bearing and clearly marked features of classic Greek regularity. The type is seldom extremely blond, and is often dark. These Latinized Vikings were and are animated by a restless and nomadic energy and by a fierce aggressiveness. They played a brilliant role during the twelfth and following centuries, but later on the continent this strain ran out. The type is still very common among the English of good families, and especially among hunters, explorers, navigators, adventurers, and officers of the lesser ranks in the British army. These latter-day Normans are natural rulers and administrators, and it is to this type that England largely owes her extraordinary ability to govern justly and firmly the lower races. This Norman blood occurs often among the native Americans, but with the changing social conditions and the filling up of the waste places of the earth, it is doomed to a speedy extinction.

The invasion of the Normans strengthened the Nordic and not the Mediterranean elements in the British Isles, but the connection once established with France, especially with Aquitaine, later introduced from southern France certain brunet elements of Mediterranean affinities.

The Nordics in England are in these days apparently receding before the little brunet Mediterranean type. The causes of this decline are the same as in France, and the chief loss is through the wastage of blood by war and emigration.

An extremely potent influence, however, is the / transformation of the nation from an agricultural / to a manufacturing community. Heavy, healthful work in the fields of northern Europe enables the Nordic type to thrive, but the cramped factory and crowded city quickly weeds him out, while the little brunet Mediterranean can work a spindle, set type, sell ribbons, or push a clerk's pen far better than the big, clumsy, and somewhat heavy Nordic blond, who needs exercise, meat, and air, and cannot live under Ghetto conditions.

The increase of urban communities at the ex\ pense of the countryside is also an important eleJ J ment in the fading of the Nordic type, because the \ energetic countryman of this blood is more apt to improve his fortunes by moving to the city than the less ambitious Mediterranean. The country villages and the farms are the nurseries of nations, while cities are consumers and seldom producers of men.

If England has deteriorated, and there are those who think they see indications of such decline, it is due to the lowering proportion of the Nordic blood and the transfer of political power from the vigorous Nordic aristocracy and middle classes to the radical and labor elements, both largely recruited from the Mediterranean type.

Only in Scandinavia and north Germany does the Nordic race seem to maintain its full vigor in spite of the enormous wastage of three thousand years of swarming forth of its best fighting men.

Holland and Flanders are purely Teutonic, the Flemings being the descendants of those Franks who did not adopt Latin speech as did their Teutonic kin across the border in Artois and Picardy; and Holland is the ancient Batavia with the Frisian coast lands eastward to old Saxony.

Denmark, Norway, and Sweden are purely Nordic and yearly contribute swarms of a splendid type of immigrants to America, and are now, as they have been for thousands of years, the nursery and broodland of the master race.

In mediæval times the Norse and Danish Vikings sailed not only the waters of the known Atlantic, but ventured westward through the fogs and frozen seas to Iceland, Greenland, and America. Sweden, after sending forth her Goths and other early Teutonic tribes, turned her attention to the shores of the eastern Baltic, colonized the coast of Finland and the Baltic provinces, and supplied as well a strong Scandinavian element to the aristocracy of Russia.

The coast of Finland is, as a result, Swedish, and the natives of the interior have distinctly Nordic characters with the exception of the skull, which in its roundness shows traces of an ancient Alpine crossing.

The population of the so-called Baltic provinces of Russia is everywhere Nordic, and their affinities are with Scandinavia and Germany rather than with Slavic Moscovy. The most primitive Aryan languages, namely, Lettish, Lithuanian, and the recently extinct Old Prussian, are found in this neighborhood, and here we are not far from the original Nordic homeland.

# IX
# THE NORDIC FATHERLAND

The area in Europe where the Nordic race developed, and in which the Aryan languages took their origin, probably included the forest region of eastern Germany, Poland, and Russia, together with the grasslands which stretched from the Ukraine eastward into the steppes south of the Ural. For reasons already explained this area was long isolated from the rest of the world, especially from Asia. When the unity of the Aryan race and of the Aryan language was broken up during the Bronze Age, the early Nordics pushed west along the sandy plains of the north and pressed against and through the Alpine populations of central Europe. They also swept down through Thrace into Greece and Asia Minor, while other large and important groups entered Asia partly through the Caucasus Mountains but in greater strength around the north and east sides of the Caspian-Aral Sea.

That portion of the Nordic race which continued to inhabit south Russia and grazed their flocks of sheep and herds of horses on the grasslands, were the Scythians of the Greeks, and from these nomad shepherds came the Cimmerians, Persians, Sacæ, Massagetæ, and perhaps the Kassites and Mitanni, and other early Aryan-speaking Nordic invaders of Asia. The descendants of these Nordics are scattered everywhere in Russia, but are now submerged by the later Slavs.

Well-marked characters of the Nordic race enable us to distinguish it definitely wherever it first appears in history, and we know that all the blondness in the world is derived from this source. When it first enters the Mediterranean world coming from the north, its arrival is everywhere marked by a new and higher civilization. In most cases the contact of the vigorous barbarians with the ancient civilizations created a sudden impulse of life and an outburst of culture as soon as the first destruction wrought by the conquest was repaired.

In addition to the long continued selection exercised by the severe climatic conditions of the north, and the consequent elimination of ineffectives, all of which affects a race, there is another force at work which concerns the individual as well. The energy developed in the north is not immediately lost when transferred to the softer conditions of existence in the Mediterranean and Indian countries. This energy endures for several generations, and only dies slowly away as the northern blood becomes diluted and the impulse to strive fades.

The contact of Hellene and Pelasgian caused the blossoming of the ancient civilization of Hellas, just as two thousand years later, when the Nordic invaders of Italy had absorbed the science, art, and literature of Rome, they produced that splendid century we call the Renaissance.

The chief men of the Cinque Cento were of Nordic, largely Gothic and Lombard, blood, a fact easily recognized by a close inspection of busts or portraits in north Italy. Dante, Raphael, Titian, Michael Angelo, Leonardo da Vinci were all of Nordic type.

Similar expansions of civilization and organization of empire, followed the incursion of the Nordic Persians into the land of the round skull Medes, and the introduction of Sanskrit into India by the Nordic Sacæ who conquered that peninsula. These outbursts of progress, due to the first contact and mixture of two contrasted races, are, however, only transitory and pass with the last lingering trace of Nordic blood.

In India the blood of these Aryan-speaking invaders has been absorbed by the dark Hindu, and in the final event only their synthetic speech survived.

The marvellous organization of the Roman state made use of the services of Nordic mercenaries, and kept the Western Empire alive for three centuries after the blood of the ancient Romans had virtually ceased to exist. The date when the population of the Empire had become predominantly of Mediterranean and Oriental blood, due to the introduction of slaves from the east and the wastage of Italian blood in war, coincides with the establishment of the Empire under Augustus, and the last Republican patriots represent the final protest of the old patrician Nordic strain. For the most part they refused to abdicate their right to rule in favor of manumitted slaves and imperial favorites, and fell in battle and sword in hand. The Roman died out but the slaves survived, and their descendants predominate among the south Italians of to-day.

The Byzantine Empire, from much the same causes, in its turn gradually became less and less European and more and more Oriental until it, too, withered away.

When these facts are considered the fall of Rome ceases to be a mystery, and the only wonder is that the Roman state lived on after the Romans were extinct, or that the Eastern Empire struggled on so long with an ever fading Greek population. Both in Rome and in Greece only the language of the dominant race survived.

So entirely had the blood of the Romans vanished in the last days of the Empire that sorry bands of barbarians wandered at will through the desolated provinces. Cæsar and his legions would have made short work of these unorganized banditti, but Caesar and his legions had become a memory, although that memory was great enough to inspire in the intruders a certain awe and desire to imitate. Against invaders, however, blood and brawn are more effective than tradition and culture, however noble these may be.

Early ascetic Christianity played a large part in this decline of the Roman Empire, as it was at the outset the religion of the slave, the meek, and the lowly, while Stoicism was the religion of the strong men of the time. This bias in favor of the weaker elements greatly interfered with their elimination by natural processes, and the fighting force of the empire was gradually undermined. Christianity was in sharp contrast to the worship of tribal deities which preceded it, and tended then, as it does now, to break down class and race distinctions. Such distinctions are absolutely essential to the maintenance of race purity in any community when two or more races five side by side.

Race feeling may be called prejudice by those whose careers are cramped by it, but it is a natural antipathy which serves to maintain the purity of type. The unfortunate fact that nearly all species of men interbreed freely leaves us no choice in the matter. Either the races must be kept apart by artificial devices of this sort, or else they ultimately amalgamate, and in the offspring the more generalized or lower type prevails.

# X
# NORDIC RACE OUTSIDE OF EUROPE

We find few traces of Nordic characters outside of Europe. When Egypt was invaded by the Libyans from the west in 1230 B. C., they were accompanied by blond "sea people," probably the Achæan Greeks, and it is interesting to note that a certain amount of reddish blondness exists today on the northern slopes of the Atlas Mountains. That it is of Nordic origin we may be certain, but through what channels it came we have no means of knowing. There is no historic invasion of north Africa by Nordics except the Vandal conquests, but there does not seem to be any probability that this small Teutonic tribe left behind it any physical trace in the native population.

The Philistines and Amorites of Palestine may have been of the Nordic race. Certain references to the size of the sons of Anak and to the fairness of David, whose mother was an Amoritish woman, point vaguely in this direction.

References in Chinese annals to the green eyes of the Wu-suns or Hiung-Nu in central Asia are the only sure evidence we have of the Nordic race in contact with the peoples of eastern Asia.

The so-called blondness of the hairy Ainus of the northern islands of Japan seems to be due to a trace of what might be called Proto-Nordic blood. The hairiness of these people is in sharp contrast to their Mongoloid neighbors, but it is a generalized character common to the highest and the lowest races of man. The primitive Australoids and the highly specialized Scandinavians are among the most hairy populations in the world. So in the Ainus this somatological peculiarity is merely the retention of a very primitive trait. The occasional brown or greenish eye, and the sometimes fair complexion of the Ainus, are, however, suggestive of Nordic affinities, and of an extreme easterly extension of Proto-Nordics at a very early period.

The skull shape of the Ainus is extremely dolichocephalic, while the broad cheek bones indicate a Mongolian cross, as in the Esquimaux. The Ainus, like many other small, mysterious people, are probably merely the remnants of one of the many early races that are fast fading into extinction. The division of man into species is very ancient, and the chief races of the earth are merely the successful survivors of the long struggle. Many species, subspecies, and races have vanished utterly, except for reversional characters which we find in the larger races.

The only Nordics in Asia Minor, so far as we know, were the Phrygians who came across the Hellespont about 1400 B. C. as part of the same migration which brought the Achaeans into Greece; the Cimmerians who entered by the same route and also through the Caucasus about 650 B. C., and still later, in 270 B. C., the Gauls who, coming from north Italy through Thrace, crossed the Hellespont and founded Galatia. So far as our present information goes, little or no trace of these invasions remains in the existing populations of Anatolia. The expansions of the Persians and the Aryanization of their empire, and the conquests of the

Nordics east and south of the Caspian-Aral Sea, will be discussed in connection with the spread of Aryan languages.

# XI
# THE RACIAL APTITUDES

Such are the three races, the Alpine, Mediterranean, and Nordic, which enter into the composition of European populations of to-day, and in various combinations comprise the great bulk of white men all over the world. These races vary intellectually and morally just as they do physically. Moral, intellectual, and spiritual attributes are as persistent as physical characters, and are transmitted unchanged from generation to generation.

In considering skull characters we must remember that, while indicative of independent descent, the size and shape of the head are not closely related to brain power. Aristotle was a Mediterranean and had a small, long skull, while Humboldt had a large and characteristically Nordic skull, but equally dolichocephalic. Socrates and Diogenes were apparently quite un-Greek and represent remnants of some early race, perhaps of Paleolithic man. The history of their lives shows clearly that each was recognized as in some degree alien by their fellow countrymen, just as the Jews apparently regarded Christ, as, in some indefinite way, un-Jewish.

Mental, spiritual, and moral traits are closely associated with the physical distinctions among the different European races, although like somatological characters, these spiritual attributes have in many cases gone astray. Enough remain, however, to show that certain races have special aptitudes for certain pursuits. The Alpine race is always and everywhere a race of peasants, an agricultural and never a maritime race. In fact, they only extend to salt water at the head of the Adriatic.

The coastal and seafaring populations of north Europe are everywhere Nordic as far as the coast of Spain, and among Europeans this race is preeminently fitted to maritime pursuits.

The Nordics are, all over the world, a race of soldiers, sailors, adventurers, and explorers, but above all, of rulers, organizers, and aristocrats in sharp contrast to the essentially peasant character of the Alpines. Chivalry and knighthood, and their still surviving but greatly impaired counterparts, are peculiarly Nordic traits, and feudalism, class distinctions, and race pride among Europeans are traceable for the most part to the north.

The mental characteristics of the Mediterranean race are well known, and this race, while inferior in bodily stamina to both the Nordic and the Alpine, is probably the superior of both, certainly of the Alpines, in intellectual attainments. In the field of art its superiority to both the other European races is unquestioned.

Before leaving this interesting subject of the correlation of spiritual and moral traits with physical characters, we may note that these influences are so deeply rooted in everyday consciousness that the average novelist or playwright would not fail to make his hero a tall, blond, honest, and somewhat stupid youth, or his villain a small, dark, and exceptionally intelligent individual of warped moral character. The gods of Olympus were almost all described as blond, and it would be difficult to imagine a Greek artist' painting .a brunette Venus. In church pictures today all angels are blonds, while the denizens of the lower regions revel in deep brunetness. Most ancient tapestries show a blond earl on horseback and a dark haired churl holding the bridle, and in depicting the crucifixion no artist hesitates to make the two thieves brunet in contrast to the blond Saviour. This latter is something more than a convention, as such quasi-authentic traditions as we have of our Lord indicate his Nordic, possibly Greek, physical and moral attributes.

These and other similar traditions clearly point to the relation of one race to another in classic, mediæval, and modern times. How far they will be modified by democratic institutions and the rule of the majority remains to be seen.

The wars of the last two thousand years in Europe have been almost exclusively wars between the various nations of this race, or between rulers of Nordic blood.

From a race point of view the present European conflict is essentially a civil war, and nearly all the officers and a large proportion of the men on both sides are members of this race. It is the same old story of mutual butchery and mutual destruction between Nordics, just as the Nordic nobility of Renaissance Italy seem to have been possessed with a blood mania to kill one another off. It is the modern edition of the old berserker blood rage, and is class suicide on a gigantic scale. It is hard to say on which side there is a preponderance of Nordic blood, as Flanders and northern France are more Teutonic than south Germany, and the backbone of the armies that England has put in the field, together with those of her colonies, are almost purely Nordic, while a large portion of the Russian armies is of the same race.

The writer has carefully refrained in this article from the use of the words "Teutonic" and "Germanic" except in their most limited sense, because the names are currently used in a national and not in a racial sense, to denote the inhabitants of the central empires. Such broader use would include millions who are totally un-Teutonic, and exclude millions, of pure Teutonic blood who are outside of the political borders of Austro-Germany.

# XII
# ARYA

Having shown the existence in Europe of three subspecies of distinct origin and a single predominant type of language called the Aryan or synthetic group, it remains to inquire to which of the three races can be assigned the honor of inventing, elaborating, and introducing this most highly developed form of human speech, and our investigations will show that the facts point indubitably toward an original unity between the Nordic, or rather the Proto-Nordic race and the Pro to-Aryan language or the generalized, ancestral, Aryan mother tongue.

Of the three claimants to the honor of being the original creator of the highest form of synthetic speech, known as the Aryan group of languages, we can at once dismiss the Mediterranean race.

The members of this race on the south shores of the Mediterranean, the Berbers and the Egyptians, speak now, and have always spoken, non- Aryan tongues. In Asia, also, many people of this race speak non-Aryan tongues. We also know that the speech of the original Pelasgians was not Aryan, that in Crete remnants of Pre-Aryan speech persisted until about 500 B. C., and that the Hellenic language was introduced into Ægean countries from the north. In Italy the Ligurian and Etruscan in the north, and the Messapian in the south, were non-Aryan languages; and the ancestral form of Latin speech in the guise of Umbrian and Oscan came through the Alps from the countries beyond.

Into Spain the Celtiberian language was introduced from the north about 600 B. C., but with so little force behind it that it was unable to entirely replace the non-Aryan language of the aborigines, which continues to this very day as Basque.

In Britain Aryan speech was introduced about 800 B. C., and in France somewhat earlier. In central and northern Europe no certain trace of nonAryan languages at one time spoken there persists, except among the Lapps and in the neighborhood of the Gulf of Finland, where the non-Aryan Finnic dialects are spoken to-day by the Finlanders and the Esthonians.

We thus know the approximate dates of the introduction of Aryan speech into western and southern Europe, and that it came in through the medium of the Nordic race. On the southern coast of the inland sea, including Egypt, the population spoke in ancient times, and still speaks, non-Aryan tongues; and in Spain and in the adjoining parts of France nearly half a million people continue to speak an agglutinative language, called Basque or Euskarian. In skull shape these Basques correspond closely with the Aryan-speaking populations around them, being dolichocephalic in Spain, and brachycephalic in France. In the case of both the long skull and the round skull Basques, the lower part of the face is long and thin with a peculiar and pointed chin. In other words, their faces show certain secondary racial characters which have been imposed by selection upon a people composed originally of two races of independent origin, but long isolated by the limitations of language.

Other than the Basque language there are in western Europe but few remnants of Pre-Aryan speech, and these are found chiefly in place names and in a few obscure words.

Remnants of non-Aryan speech exist here and there throughout European Russia, but many of them can be traced to historic invasions. Until we reach the main body of Ural-Altaic speech in the east of Russia, the Esths, with kindred but small tribes of Livonians and Tchouds, and the Finns alone can lay claim to the honor of antedating the Aryan tongue in Moscovite territories, but the physical type of all these tribes is distinctly Nordic. In this connection the Lapps and related groups in the far north can be disregarded.

The problem of the Finns is a difficult one. The coast of Finland, of course, is purely Swedish, but the great bulk of the population in the interior brachycephalic, though otherwise thoroughly Nordic in type. It would seem that here the Alpine element were the more ancient.

The most important non-Aryan language in Europe is the Magyar of

Hungary, but this we know was introduced from the eastward at the end of the ninth century.

In the Balkans the language of the Turks has never been a vernacular as it is in Asia Minor. In Europe it was spoken only by the soldiers and the civil administrators, and by very sparse colonies of Turkish settlers. The mania of the Turks for white women, which is said to have been one of the -motives that led to the conquest of the Byzantine Empire, has unconsciously resulted in the obliteration of the Mongoloid type of the original Asiatic invaders. Persistent crossing with Circassian and Georgian women, as well as with slaves of every race in Asia Minor or in Europe with whom they came in contact, has made the European Turk of to-day indistinguishable in physical characters from his Christian neighbors.

The Turks of Seljukian and Osmanli origin were never numerous, and the Sultan's armies were and are largely composed of Islamized Anatolians and Europeans.

In Persia and India, also, the Aryan languages were introduced from the north at known periods, so in view of all these facts, the Mediterranean race cannot claim the honor of either the invention or dissemination of the synthetic languages.

The chief claim of the Alpine race of central Europe and western Asia to the invention and introduction into Europe of the Proto-Aryan form of speech rests on the fact that nearly all the members of this race in Europe speak well developed forms of Aryan speech, chiefly in the form of Slavic. This fact taken by itself may have no more significance than the fact that the Mediterranean race in Spain, Italy, and France speaks Romance languages, but it is, nevertheless, an argument of some weight.

Outside of Europe the Armenians and other Armenoid brachycephalic peoples of Asia Minor and the Iranian Highlands, all of Alpine race, together with a few isolated tribes of the Caucasus, speak Aryan languages, and these peoples lie on the highroad along which knowledge of the metals and other cultural developments entered Europe.

If the Aryan language were invented and developed by these Armenoid Alpines we should be obliged to assume that they introduced it along with bronze culture into Europe about 3000 B. C. and taught the Nordic blonds both their language and their metal culture. There are, however, in western Asia many Alpine peoples who do not speak Aryan languages and yet are Alpine in type, such as the Turcomans, and in Asia Minor the so-called Turks are also largely Islamized Alpines of the Armenoid subspecies who speak Turki. There is no trace of Aryan speech south of the Caucasus until after 1700 B. C., and the Hittite language spoken before that date in central and eastern Asia Minor, although not yet clearly deciphered, was non-Aryan to the best of our present knowledge. The Hittites themselves were probably ancestral to the living Armenians.

We are thoroughly acquainted with the languages of all the Mesopotamian countries, and we know that the speech of Accad and Sumer, of Susa and Media was agglutinative, and that the languages of Assyria and of Palestine were Semitic. The speech of the Kassites was Aryan, and the language of the shortlived empire of the Mitanni in the foothills south of Armenia, is the only one about the character of which there can be some doubt, but in all probability it was Aryan. There is, therefore, much negative evidence against the existence of Aryan speech in this part of the world earlier than its known introduction by Nordics.

If the last great expansion into Europe of the Alpine race brought from Asia the Aryan mother tongue, as well as the knowledge of metals, we must assume that all the members of the Nordic race thereupon adopted synthetic speech from the Alpines.

We know that these Alpines reached Britain about 1800 B. C., and probably had previously occupied much of Gaul, so that if they are to be credited with the introduction of the synthetic languages into western Europe, it is difficult to understand why we have no known trace of any form of Aryan speech in central Europe or west of the Rhine prior to 1000 B.C., while we have some, though scant, evidence of non-Aryan languages.

Even assuming, however, that the Alpines did introduce this synthetic language to the Baltic dolichocephs along with the art of metallurgy, we are obliged to believe that the Nordics, equipped with this synthetic language and with bronze weapons, starting on their marvellous career of expansion a full millennium after the Alpine conquest, first attacked and conquered their Alpine teachers and then poured down from the north in successive waves into the domain of the Mediterranean race, passing en route through brachycephalic countries and taking along with them varying proportions of Alpine blood.

It may be said in favor of this claim of the Alpine race to be the original inventors of synthetic speech, that language is ever a measure of culture, and the higher forms of civilization are greatly hampered by the limitations of speech imposed by the less highly evolved languages, namely, the monosyllabic and the agglutinative, which include nearly all the non- Aryan languages of the world.

It does not seem probable that barbarians, however fine in physical type and however well endowed with the potentiality of intellectual and moral development, dwelling as hunters in the bleak and barren north along the edge of the retreating glaciers and as nomad shepherds in the Russian grasslands, could have evolved a more complicated and higher form of articulate speech than the inhabitants of southwest Asia, who many thousand years earlier were highly civilized and are known to have invented the arts of agriculture, metal working and domestication of animals, as well as of writing and pottery. Nevertheless, such ' seems to be the fact.

To conclude then, a study of the Mediterranean race shows that, so far from being purely European, it is equally African and Asiatic, and in the narrow coastal fringe of southern Persia, in India, and even farther east the last strains of this race gradually fade into the negroids through prolonged cross breeding, and a similar inquiry into the origin and distribution of the Alpine species shows clearly the fundamentally Asiatic origin of this type, and that on its easternmost borders in central Asia it marches on the round skulled Mongolian.

# XIII
# THE ORIGIN OF THE ARYAN LANGUAGES

By the process of elimination set forth in the preceding chapter we are compelled to consider that the strongest claimant for the honor of being the race of the original Aryans, is the tall, blond Nordic. A study of the various languages of the Aryan group reveals an extreme diversity which can be best explained by the hypothesis that the existing languages are now spoken by people upon whom Aryan speech has been forced from without. This theory corresponds exactly with the known historic fact that the Aryan languages, during the last three or four thousand years at least, have, again and again, been imposed by Nordics upon populations of Alpine and Mediterranean blood.

Within the present distributional area of the Nordic race, and in the very middle of a typical area of isolation, is the most generalized member of the Aryan group, namely, Lettish, or old Lithuanian, situated on the Gulf of Riga, and almost Proto-Aryan in character. Close at hand was the closely related Old Prussian or Borussian, very recently extinct. These archaic languages are relatively close to Sanskrit, and are located in actual contact with the non-Aryan speech of the Esths and Finns.

The non-Aryan languages in eastern Russia are Ugrian, a form of speech which extends far into Asia, and which alone of all agglutinative tongues, contains elements which unite it with synthetic speech, and which is consequently dimly transitory in character. In other words, in the opinion of many philologists, a primitive form of Ugrian might have given birth to the Proto-Aryan ancestor of existing synthetic languages.

This hypothesis, if sustained by further study, will provide additional evidence that the site of the development of the Aryan languages, and of the Nordic species, was in eastern Europe, and in a region which is close to the place of contact between the most archaic synthetic languages and the most nearly related non-Aryan tongue, the agglutinative Ugrian.

The Aryan tongue was introduced into Greece by the Achaeans about 1400 B. C., and later, about 1100 B. C., by the true Hellenes, who brought in the classic dialects of Dorian, Ionian, and Æolian.

These Aryan languages superseded their non-Aryan predecessor, the Pelasgian. From the language of these early invaders came the Illyrian, Thracian, Albanian, classic Greek, and the debased modern Romaic, a descendant of the Ionian dialect.

Aryan speech was introduced among the non-Aryan Etruscans of the Italian Peninsula by the Umbrians and Oscans about 1100 B. C. These languages were ultimately succeeded by Latin, an offshoot of these early Aryan tongues of northern Italy which later spread to the uttermost confines of the Roman Empire. Its descendants to-day are the Romance tongues spoken within the ancient imperial boundaries, the Portuguese on the west, Castilian, Catalan, Provencal, French, the langue d'oïl of the Walloons, Ligurian, Romansch, Ladin, Friulian, Tuscan, Calabrian, and Rumanian.

The problem of the existence of a language, the Rumanian, in the eastern Carpathians, cut off by Slavic and Magyar tongues from the nearest Romance languages, but nevertheless clearly descended from Latin, presents great difficulties. The Rumanians themselves make two claims; the first, which can be safely disregarded, is an unbroken linguistic descent from a group of Aryan languages which occupied this whole section of Europe, from which Latin was derived, and of which Albanian is also a remnant.

The more serious claim, however, made by the Rumanians, is to linguistic and racial descent from the military colonists planted by the Emperor Trajan in the great Dacian plain. This may be possible, so far as the language is concerned, but there are some weighty objections to it.

We have no evidence for, and much against, the existence of Rumanian speech north of the Danube for nearly a thousand years after Rome abandoned this outlying region. Dacia was one of the last provinces to be occupied by Rome, and was the first from which the legions were withdrawn upon the dissolution of the empire. The northern Carpathians, furthermore, where the Rumanians claim to have taken refuge during the barbarian invasions, form part of the Slavic homeland, and it was in these same mountains, and in the Ruthenian districts of eastern Galicia, that the Slavic languages were developed, probably by the Sarmatians and Venethi, and from which they spread in all directions in the centuries that immediately follow the fall of Rome. So it is almost impossible to credit the survival of a frontier community of Romanized natives situated not only in the path of the great invasions of Europe from the east, but also in the very spot where Slavic languages were at the time evolving.

Rumanian speech occupies a large area outside of the present kingdom of Rumania, in Russian Bessarabia, Austrian Bukowina, and above all in Hungarian Transylvania, all of which were parts of ancient Dacia, and which are now to be "redeemed" by the Rumanians.

This linguistic problem is further complicated by the existence in the Pindus Mountains of Thessaly of another large community of Vlachs of Rumanian speech. How this later community also could have survived from Roman times until to-day, untouched either by the Greek language of the Byzantine Empire or by the Turkish conquest, is another difficult problem. The solution of these questions receives no assistance from anthropology, as these Rumanian-speaking populations, both on the Danube and in the Pindus

Mountains, in no way differ physically from their neighbors on all sides. Through whatever channel they acquired their Latin speech, the Rumanians to-day can lay no valid claim to blood descent, even in a very remote degree, from the true Romans.

The first Aryan languages known in western Europe were the Celtic group which first appears west of the Rhine about 1000 B. C.

There have been found only a few dim traces of Pre-Aryan speech in the British Isles, these chiefly in place names. In Britain Celtic speech was introduced in two successive waves, first by the Goidels, or "Q Celts," who apparently appeared about 800 B. C., and this form exists to this day as Erse in western Ireland, as Manx of

the Isle of Man, and as Gaelic in the Scottish Highlands.

The Goidels were of bronze culture. When they reached Britain they must have found there a population preponderantly of Mediterranean type with numerous remains of still earlier races of Paleolithic times, and also some round skull Alpines of the Round Barrows, who have since faded from the living population. When the next invasion, the Cymric, occurred, the Goidels had been very largely absorbed by these underlying Mediterranean aborigines who had accepted the Goidelic form of Celtic speech, just as on the continent the Gauls had mixed with Alpine and Mediterranean natives though imposing upon the conquered their own tongue. In fact, in Britain, Gaul, and Spain the , Goidels and Gauls were chiefly a ruling, military class, while the great bulk of the population remained unchanged, although Aryanized in speech.

The Brythonic or Cymric tribes, or "P Celts," followed about five hundred years later, driving the Goidels westward through Germany, Gaul, and Britain, as is proved by the distribution of place names, and this movement of population was still going on when Caesar crossed the Channel. The Brythonic group gave rise to the modern Cornish, extinct within a century, the Cymric of Wales, and the Armorican of Brittany.

In central Europe we find traces of these same two forms of Celtic speech, with the Goidelic everywhere the older and the Cymric the more recent arrival.

When the two Celtic-speaking races came into conflict in Britain their original relationship had been greatly obscured by the crossing of the Goidels with the underlying dark Mediterranean race of Neolithic culture, and by the mixture of the Belgæ with Teutons. The result of all this was that the Brythons did not distinguish between the blond Goidels and the brunet, but Celticized Mediterraneans, as they all spoke Goidelic dialects.

In the same way when the Teutonic tribes entered Britain they found there peoples all speaking Celtic of some form, either Goidelic or Cymric, and promptly called them all Welsh (foreigners). These Welsh were preponderantly of Mediterranean type with some mixture of a blond Goidel strain and a much stronger blond strain of Cymric origin, and these same elements exist to-day in England. The Mediterranean race is easily distinguished, but the physical types derived from Goidel and Brython alike are merged and lost in the later floods of pure Nordic blood, Angle, Saxon, Dane, Norse, and Norman. In this primitive, dark population, with successive layers of blond Nordics imposed upon it, each one more purely Nordic, lies the secret and the solution of the anthropology of the British Isles. This Iberian substratum was able to absorb, to a large extent, the earlier Celtic-speaking invaders, both Goidels and Brythons, but it is only just beginning to seriously threaten the Teutonic Nordics, and to reassert its ancient brunet characters after three thousand years of submergence.

In northwest Scotland there is a Gaelic-speaking area where the place names are all Scandinavian, and the physical types purely Nordic. This is the only spot in the British Isles where Celtic speech has re-conquered a district from the Teutonic languages, and it was the site of one of the earliest conquests of the Norse Vikings, probably in the early centuries of our era. In Caithness in north Scotland, as well as in some isolated spots .on the Irish coasts, the language of these same Norse pirates persisted until within a century. In the fifth century of our era and after the breakup of Roman domination in Britain there was much racial unrest, and a back wave of Goidels crossed from Ireland and either introduced or reinforced the Gaelic speech in the highlands. Later, Goidelic speech was gradually driven north and west by the intrusive English of the lowlands, and was ultimately forced over this originally Norse-speaking area.

We have elsewhere in Europe evidence of similar shiftings of speech without corresponding changes in the blood of the population.

Except in the British Isles and in Brittany, Celtic languages have left no modern descendants, but have everywhere been replaced by languages of Neo-Latin or Teutonic origin. Outside of Brittany one of the last, if not quite the last, references to Celtic speech in Gaul is the historic statement that "Celtic" tribes, as well as "Armoricans," took part at Châlons in the great victory in 451 A. D. over Attila, the Hun, and his confederacy of subject nations.

On the continent the only existing populations of Celtic speech are the primitive inhabitants of central Brittany, a population noted for their religious fanaticism and for other characteristics of a backward people. This Celtic speech is said to have been introduced in the early century of our era by Britons fleeing from the Saxons. These refugees, if there were such, must have been dolichocephs of either Mediterranean or Nordic race, or both. We are asked by this tradition to believe that the skull shape of these Britons was lost, but that their language was adopted by the Alpine population of Armorica. It is much more probable that the Cymric-speaking Alpines of Brittany have merely retained in this isolated corner of France a form of Celtic speech which was prevalent throughout northern Gaul and Britain before these provinces were conquered by Rome and Latinized. Cæsar remarked that there was little difference between the speech of the Belgæ in northern Gaul and in Britain. In both cases the speech was Cymric.

Long after the conquest of Gaul by the Goths and Franks, Teutonic speech was predominant among the ruling classes, and by the time it succumbed to the Latin tongue of the Romanized natives, the old Celtic languages had been entirely forgotten outside of Brittany.

An example of similar changes of language is to be found in Normandy where the country was originally inhabited by the Nordic Belgæ, who spoke a Cymric language before that tongue was replaced by Latin. This coast was ravaged about 300 or 400 A. D. by Saxons who formed settlements along both sides of the Channel and the coasts of Brittany, which were later known as the Litus Saxonicum. Their progress can best be traced by place names, as our historic record of these raids is scanty.

The Normans landed in Normandy in the year 911 A. D. They were heathen Danish barbarians, speaking a Teutonic language. The religion, culture, and language of the old Romanized populations worked a miracle in the transformation of everything except blood in one short century. So quick was the change that 155 years later the descendants of the same Normans landed in England as Christian Frenchmen, armed with all the culture of their period. The change was startling, but the blood of the Norman breed remained unchanged and entered England as a purely Nordic type.

# XIV
# THE ARYAN LANGUAGE IN ASIA

In the Ægean region and south of the Caucasus the Nordics appear after 1700 B. C., but there were unquestionably invasions and raids from the north for many centuries previous to our first records. These early migrations probably were not in sufficient force to modify the blood of the autochthonous races or to substitute Aryan languages for the ancient Mediterranean and Asiatic tongues.

These men of the North came from the grasslands of Russia in successive waves, and among the first of whom we have fairly clear knowledge were the Achaeans and Phrygians. Aryan invaders are mentioned in the dim chronicles of the Mesopotamian empires about 1700 B. C., as Kassites, and later as Mitanni. Aryan names of prisoners captured beyond the mountains in the north, and of Aryan deities before whom oaths were taken, are recorded about 1400 B. C., but one of the first definite accounts of Nordics south of the Caucasus describes the presence of Nordic Persians at Lake Urmia about 900 B. C. There were many incursions from that time on, the Cimmerians raiding across the Caucasus as early as 680 B. C., and shortly afterward overrunning all Asia Minor.

The easterly extension of the Russian steppes north of the Caspian-Aral Sea in Turkestan, as far as the foothills of the Pamirs, was occupied by the Sacæ or Massagetæ, who were also Nordics and akin to the Cimmerians and Persians. For several centuries groups of Nordics drifted as nomad shepherds across the Caucasus into the empire of the Medes, introducing little by little the Aryan tongue, which later developed into Old Persian.

In 538 B. C. these Persians had become sufficiently numerous to overthrow their rulers, and under the leadership of the great Cyrus they organized the Persian Empire, one of the most enduring of Oriental states. The base of the population of the Persian Empire rested on the round skull Medes who belonged to the Armenoid subdivision of the Alpines. Under the leadership of their priestly caste of Magi, these Medes rebelled again and again against their Nordic masters before the two peoples became fused.

From 525 to 485 B. C., during the reign of Darius, whose sculptured portraits show a man of pure Nordic type, the tall, blond Persians had become almost exclusively a class of great ruling nobles, and had forgotten the simplicity of their shepherd ancestors. Their language belonged to the Eastern or Iranian division of Aryan speech, and was known as Old Persian, which continued to be spoken until the fourth century before our era. From it were derived Pehlevi, or Parthian, and modern Persian. The great book of the old Persians, the Avesta, which was written in Zendic, also an Iranian language, does not go back to the reign of Darius, and was remodelled

after our era, but the Old Persian of Darius was closely related to the Zendic of Bactria, and to the Sanskrit of Hindustan. From Zendic, also called Medic, are derived Ghalcha, Balochi, Kurdish, and other dialects.

The rise to imperial power of the dolichocephalic Aryan-speaking Persians was largely due to the genius of their leaders, but the Aryanization of western Asia by them is one of the most amazing events in history. The whole region became completely transformed so far as the acceptance by the conquered of the language and religion of the Persians was concerned, but the blood of the Nordic race quickly became diluted, and a few centuries later disappears from history.

At the time of the great wars with Greece the pure Persian blood was still unimpaired and in control, and in the literature of the time there is little evidence of race antagonism between the Greek and the Persian leaders, although their rival cultures were sharply contrasted. In the time of Alexander the Great the pure Persian blood was obviously confined to the nobles, and it was the policy of Alexander to Hellenize the Persians and to amalgamate his Greeks with them. The amount of pure Macedonian blood was not sufficient to reinforce the Nordic strain of the Persians, and the net result was the entire loss of the Greek stock.

It is a question whether the Armenians of Asia Minor derived their Aryan speech from this invasion of the Nordic Persians, or whether they received it at an earlier date from the Phrygians, and from the west. These Phrygians entered Asia Minor by way of the Dardanelles and broke up the Hittite Empire. Their language was Aryan, and probably related to Thracian. In favor of the theory of the introduction of the Armenian language by the Phrygians from the west, rather than by the Persians from the east, is the highly significant fact that the basic structure of that tongue shows its relationship to be with the western rather than with the eastern group of Aryan languages, and this, too, in spite of a very large Persian vocabulary.

The Armenians themselves, like all the other natives of the plateaux and highlands as far east as the Hindu Kush Mountains, while of Aryan speech, are of the Armenoid subdivision, in sharp contrast to the predominant types south of the mountains in Persia, Afghanistan, and Hindustan, all of which are dolichocephalic and of Mediterranean affinity, but generally betraying traces of admixture with still more ancient races of negroid origin, especially in India.

We now come to the last and easternmost extension of Aryan languages in Asia. As mentioned above, the grasslands and steppes of Russia extend north of the Caucasus Mountains and the Caspian Sea to ancient Bactria, now Turkestan. This whole country was occupied by the Nordic Sacse and the closely related Massagetae. At a very early date, probably about the beginning of the second millennium B. C., or perhaps even earlier, the first Nordics crossed over the Afghan passes, entered the plains of India, and organized a state in the Penjab, "the land of the five rivers," bringing with them Aryan speech among a population probably of Mediterranean type, and represented to-day by the Dravidians. The Nordic Sacse arrived later in India and introduced the Vedas, religious poems, which were at first transmitted orally, and which were reduced to written form in Old Sanskrit by the Brahmans at the comparatively late date of 300 A. D. From this classic Sanskrit are derived all the modern Aryan languages of Hindustan, as well as the Singalese of Ceylon and the chief dialects of Assam.

There is great diversity of opinion as to the date of the first entry of these Aryan-speaking tribes into the Penjab, and the consensus of opinion seems to indicate a period between 1600 and 1700 B. C. or even somewhat earlier. However, the very close affinity of Sanskrit to the Old Persian of Darius and to the Zendavesta would strongly indicate that the final introduction of Aryan languages in the form of Sanskrit occurred at a much later date.

If close relationship between languages indicates correlation in time, then the entry of the Sacæ into India would appear to have been nearly simultaneous with the crossing of the Caucasus by the Nordic Cimmerians and their Persian successors.

The relationship between the Zendavesta and the Sanskrit Vedas is as near as that between High and Low German, and consequently such close affinity prevents our thrusting back the date of the separation of the Persians and the Sacæ more than a few centuries.

A simultaneous migration southward of nomad shepherds on both sides of the Caspian-Aral Sea would naturally occur in a general movement southward, and such migrations may have taken place several times. In all probability these Nordic invasions occurred one after another for a thousand years or more, the later ones obscuring and blurring the memory of their predecessors.

When shepherd tribes leave their grasslands and attack their agricultural neighbors, the reason is nearly always famine due to prolonged drought, and causes such as these have again and again in history put the nomad tribes in motion over large areas. During many centuries fresh tribes composed of Nordics, or under the leadership of Nordics, but all Aryan-speaking, poured over the Afghan passes from the northwest and pushed before them the earlier arrivals. Clear traces of these successive floods of conquerors are to be found in the Vedas themselves.

The Sacæ and Massagetae were, like the Persians, blond dolichocephs, and they have left behind them dim traces of their blood among the living, Mongolized nomads of Turkestan, the Kirghizes. Ancient Bactria maintained its Nordic and Aryan aspect long after Alexander's time, and did not become Mongolized and receive the sinister name of Turkestan until the seventh century, when it was the first victim of the great series of ferocious invasions from the north and east, which, under various Mongol leaders, destroyed civilization in Asia and threatened its existence in Europe. These tall, blue eyed, Aryan-speaking Sacæ were the most easterly members of the Nordic race of whom we have record. The Chinese knew well these "green eyed devils," whom they called by their Tatar name, the "Wu-suns," the tall ones, and with whom they came into contact in about 200 B. C. in what is now Chinese Turkestan.

The Zendic form of the Iranian group of Aryan languages continued to be spoken by these Sacæ who remained in old Bactria, and from it is derived a whole group of closely related dialects still spoken in the Pamirs, of which Ghalcha is the best known.

The most easterly known Aryan tongue has been recently discovered in Turkestan. It is called Tokharian, and is undoubtedly a pure Aryan language, related, curiously enough, to the western group rather than to the Indo-Iranian. It has been deciphered from recently found inscriptions, and was a living language prior to the ninth century A. D. This constitutes another proof of the extent and duration of the Nordic occupation of Bactria.

Of all the wonderful conquests of the Sacæ there remain as evidence of their invasions only these Indian and Afghan languages. Dim traces of their blood, as stated before, have been found in the Pamirs and in Afghanistan, but in the south their blond traits have vanished, even from the Penjab. It may be that the stature of some of the hill tribes and of the Sikhs, and some of the facial characters of the latter, are derived from this source, but all blondness of skin, hair, or eye of the original Sacæ have utterly vanished.

The long skulls all through India are to be attributed to the Mediterranean race rather than to this Nordic invasion, while the Pre-Dravidians and negroids of south India, with which the former are largely mixed, are also dolichocephs.

In short, the introduction in Iran and India of Aryan languages, Iranian, Ghalchic, and Sanskrit, represents a linguistic and not an ethnic conquest.

In concluding this revision of the racial foundations upon which the history of Europe has been based, it is scarcely necessary to point out that the actual results of the spectacular conquests and invasions of history have been far less permanent than those of the more insidious victories arising from the crossing of two diverse races, and that in such mixtures the relative prepotency of the various human subspecies in Europe appears to be in inverse ratio to their social value.

The continuity of physical traits and the limitation of the effects of environment to the individual only are now so thoroughly recognized by scientists that it is at most a question of time when the social consequences which result from such crossings will be generally understood by the public at large. As soon as the true bearing and import of the facts are appreciated by lawmakers, a complete change in our political structure will inevitably occur, and our present reliance on the influences of education will be superseded by a readjustment based on racial values.

Bearing in mind the extreme antiquity of physical and spiritual characters and the persistency with which they outlive those elements of environment termed language, nationality, and forms of government, we must consider the relation of these facts to the development of the race in America. We may be certain that the progress of evolution is in full operation to-day under those laws of nature which control it, and that the only sure guide to the future lies in the study of the operation of these laws in the past.

We Americans must realize that the altruistic ideals which have controlled our social development during the past century, and the maudlin sentimentalism that has made America "an asylum for the oppressed," are sweeping the nation toward a racial abyss. If the Melting Pot is allowed to boil without control, and we continue to follow our national motto and deliberately blind ourselves to all "distinctions of race, creed, or color," the type of native American of Colonial descent will become as extinct as the Athenian of the age of Pericles, and the Viking of the days of Rollo.

# BIBLIOGRAPHY

The following list of works will be of assistance to such readers as may desire to investigate the aspects of anthropology treated in this book.


Avebury, Lord:
Prehistoric Times. 1913.

Beddoe, J.:
Various writings.

Boule, M.:
Revue d'Anthropologie. 1888, iQO A , and 1908. Breuil, l'Abbe H.:
Various writings.

Broca, Paul:
Various writings.

Cartailhac, E.:
Various writings.

Chamberlain, Houston Stewart:
Foundations of the XIXth Century.

Collignon, R.:
Various writings.

Darwin, Charles:
Descent of Man.

Davenport, Charles Benedict:
Heredity in Relation to Eugenics. 1911.

Deniker, J.:
The Races of Man. 1901.

Duckworth, W. L. H.:
Morphology and Anthropology. 1904.
Prehistoric Man. 1912.

Flinders-Petrie, W. M.:
Revolutions of Civilization. 1912.

Galton, Sir Francis:
Hereditary Genius. 1892.

Gowland, W.:
Metals in Antiquity. Jour. Roy. Anth. Inst., XLII, 1912, p. 245 et seq.

Haddon, A. C.:
Wanderings of Peoples. 1912.
Races of Man. •
The Study of Man. 1898.

Harlé, E.:
Various writings.

Hauser, O.:
Various writings.
Hrdlička, Dr. A.:
The Most Ancient Skeletal Remains of Man. 1914. Huntington, Ellsworth: Pulse of Asia. 1907.
Palestine and Its Transformation. 1911. Civilization and Climate. 1915.
Johnston, Sir Harry H.:
Views and Reviews. 1912.
Colonization of Africa. 1905.
The Opening Up of Africa. 1911.
Keane, A. H.:
Man, Past and Present. 1900.
Ethnology. 1901.
Keith, Arthur:
Antiquity of Man. 1915.
Klaatsch, H.:
Homo Aurignacius Hauseri. 1909.
Klaatsch, H., and Hauser, O.:
Archiv fur Anthropologie. 1908.
MacCurdy, G. G.:
The Eolithic Problem. 1905.
The Antiquity of Man in Europe. 1910. Metchnikoff, Elie: Nature of Man. 1903.
Mierow, Chas. C.:
The Gothic History of Jordanes.
Morgan, Thomas Hunt:
Heredity and Sex. 1914.
Heredity and Environment. 1915.
Munro, John:
Story of the British Race. 1907.
Munro, R.:
Paleolithic Man and the Terramara Settlements. Obermaier, H.: L'Anthropologie. 1908 and 1909.
Osborn, Henry Fairfield:
Age of Mammals. 1910.
Men of the Old Stone Age. 1915. •
Payne, Edward John:
History of the New World Called America. 1899.
Penck, A.:
Zeitschrift fur Ethnologie. 1908.
Peyrony, M., and Capitan:
Bulletins de la Societe d'Anthropologie de Paris. 1909-1910.

Quatrefages, A. de:
Various writings.
Rathgen, F.:
Die Metalle im Alterthum. 1915.
Reid, G. Archdall:
Principles of Heredity. 1905.
Laws of Heredity. 1910.
Retzius, A. A.:
Various writings.
Retzius, M. G.:
Various writings.
Ridgeway, Wm.:
Early Age in Greece. 1907.
The Thoroughbred Horse. 1905.
Ripley, W. Z.:
Races of Europe. 1899.
Rutot, A.:
Various writings.
Salisbury, R. D., and Chamberlain, T. C.:
Geology. 1905.
Schoetensack, 0.:
Der Unterkiefer des Homo heidelbergensis. 1908. Schwalbe, G.:
Vorgeschichte des Menschen. Zeitschrift fur Morphologic und Anthropologie. 1906.
Sergi, G.:
The Mediterranean Race. 1901.
Smith, G. Elliot:
The Ancient Egyptians. 1911. And other writings. Sollas, W. J.:
Ancient Hunters. 1911.
Taylor, Isaac:
Various writings.
Villari, Pasquale:
The Barbarian Invasions of Italy. 1902.
Woodruff, Charles Edward:
Effects of Tropical Light on White Men. 1905. Expansion of Races. 1909.
Woods, Frederick Adams:
Heredity in Royalty. 1906.
Woodward, A. S.:
Various writings.
Zaborowski, S.:
Paris. Les Aryens en l'Asie et l'Europe.

## PROVISIONAL OUTLINE OF NORDIC I

| B C. | Great Britain | Scandinavia | Germany and Austria | France and Spain | Italy |
| --- | --- | --- | --- | --- | --- |
| Before 3000 | | | Copper. | | Copper. |
| 3000–2500 | | | Great expansion of Alpines, introducing bronze into Austria and later into Germany. | Copper. | Great expansion of Alpines, intro bronze in north Italy, 3000–280 Similar invasion in south from . |
| 2500–1800 | Neolithic. Copper. Alpine invasion with bronze culture. | Neolithic. Alpine invasion with bronze culture reaches Denmark and southwest Norway. | | Alpine invasion with the bronze culture in France. | Eneolithic culture. |
| 1800–1600 | Round barrows. Megaliths. | | | Later same wave of invasion enters Spain. Megaliths. | |
| 1600–1400 | | | Hallstatt iron culture in Austrian Tyrol. | | |
| 1400–1200 | | | | | |
| 1200–1000 | | Nordic Teutons cross from Scandinavia to south coasts of Baltic and to Denmark. | Nordic Goidels in occupation as far west as the Elbe. First invasion of Nordic Teutons from Scandinavia. | 1000. Nordic Goidels cross Rhine and introduce Aryan speech (Gaulish). | c. 1100. Nordic Umbrians and introduce first Aryan speech northeast. Iron in Etruria, 1100. |
| 1000–800 | First Nordics—Goidels. First iron swords, 800. | | Goidels driven south and west by Cymty. | Hallstatt iron culture. | |
| 800–600 | First Aryan speech. | | Goidels expelled from Germany by Cymry. Expansion of Teutons in Germany. | Nordic Goidels cross Pyrenees and introduce Aryan speech in Spain. | Expansion of Mediterranean Etr over Umbrians to Alps. First Greek colonies in south Magna Græcia. |
| 600–400 | First Goidels in Ireland, 600. La Tène iron. | La Tène iron. | Cymric Belgæ driven westward by Teutons. La Tène iron culture. | La Tène iron culture in France. | Nordic Gauls in valley of Po— pine Gaul. |
| 400–300 | | Great expansion of Nordic Teutons out of Scandinavia. | Expansion of Teutons and expulsion of Cymry as far west as the Weser. | La Tène iron in Spain. Cymric Belgæ conquer northern France. | Gauls under Brennus sack Rom and destroy Etruria. New in of Nordics into Cisalpine Gaul. |
| 300–200 | Cymric Belgæ—invasion, c. 300. | | c. 250. First Teutons in Austria. | | Expansion of Rome. Punic Wars, 264–146. |
| 200–100 | Few Cymry in Ireland. | | Teutons drive Cymry out of Germany. Teutons cross the Rhine. | Teutons enter France. Marius destroys Teutones and Cimbri, 100 B. C. | |
| 100 to Christian Era | 55. Julius Cæsar. | | | Cæsar conquers Gaul, 59–51. | Augustus and the organization Empire. Extinction of old Romans. |
| | | | Defeat of Varus and Roman legions in old Saxony, 9 A. D. | | |

# SIONS AND METAL CULTURES

| Russia, Greece, and Balkans | Asia Minor | North Africa and Egypt | Mesopotamia and Persia | India and China |
| --- | --- | --- | --- | --- |
| 4000 B. C. Commencement of early Minoan in Crete.<br>Copper. | Alpines.<br>Introduction of bronze from Egypt. | Copper for ornaments, 5000.<br>Copper systematically mined, 3800. | Copper for ornaments, 5000. | |
| | | Discovery of bronze. | | |
| Great expansion of Alpines, introducing bronze from Asia Minor. | Bronze smelting.<br>Copper in Cyprus. | Giza skulls (Alpine), 3000. | | |
| Middle Minoan in Crete, 2000–1800.<br>Second city of Hissarlik—iron—2000. | | Period of depression with invasions from desert.<br>Phœnicians flourish. | | First Nordics enter India. |
| Early Nordic invasions.<br>Mycenæan culture. | | Hyksos in Egypt, 1700. | Invasion of Nordic Kassites and Mitanni. | Nordic states in Penjab. |
| Late Minoan in Crete, 1600–1450. | First Aryan names of deities—Cappadocia. | | | Nordic invasion. |
| Last Minoan, 1450–1200. Cnossos.<br>Mycenæan culture.<br>Nordic Achæans from south Russia introduce Aryan speech, 1400–1300. Hallstatt iron.<br>1200. Transition from bronze to iron in Crete. | Nordic Phrygians (Trojan leaders). | Hittites invade Syria.<br>1210. Blond sea people (Achæans) attack Egypt. | | |
| Trojan War, 1194–1184.<br>Nordic Hellenes—Dorians—enter Greece, 1100.<br>Iron in full development. | Hittites. | | Beginning of the greatness of Babylon, 1100. | Nordic Sacæ introduce Sanskrit. |
| | | | Nordic Persians recorded at Lake Urmia, 900.<br>Iron mines at Carchemish. | |
| Iron age in Russia.<br>Megarian colonization, 750. | Early Nordic raids.<br>Cimmerians, 680. | | Invasion of Nordic Scythians. | Nordic Hiung-Nu, beyond the Wall in China, become restless. |
| 500. End of non-Aryan speech in Crete.<br>Persian Wars, 500–440. | | Persian conquest, 525. | Nordic Persians overthrow Medes, 538.<br>Reign of Darius, 525–485. | |
| Macedon conquers Greece, 338.<br>Alexander the Great, 356–323. | | Alexander conquers Egypt, 332. | Conquests of Alexander. | Conquests of Alexander. |
| Decline of Nordic Scythians in Russia, and appearance in Russia of Alpine Sarmatians.<br>Gauls in Russia—forming Celto-Scyths. Nordic Galatians enter Thrace and Greece—Delphi, 279; cross into Asia Minor and found Galatia. | Nordic Galatians, 270. | | | Wu-Suns in Chinese Turkestan. |
| | | | | |
| | | | | |
| Alpine Sarmatians appear in Danube valley, 50 A. D. | | | | |

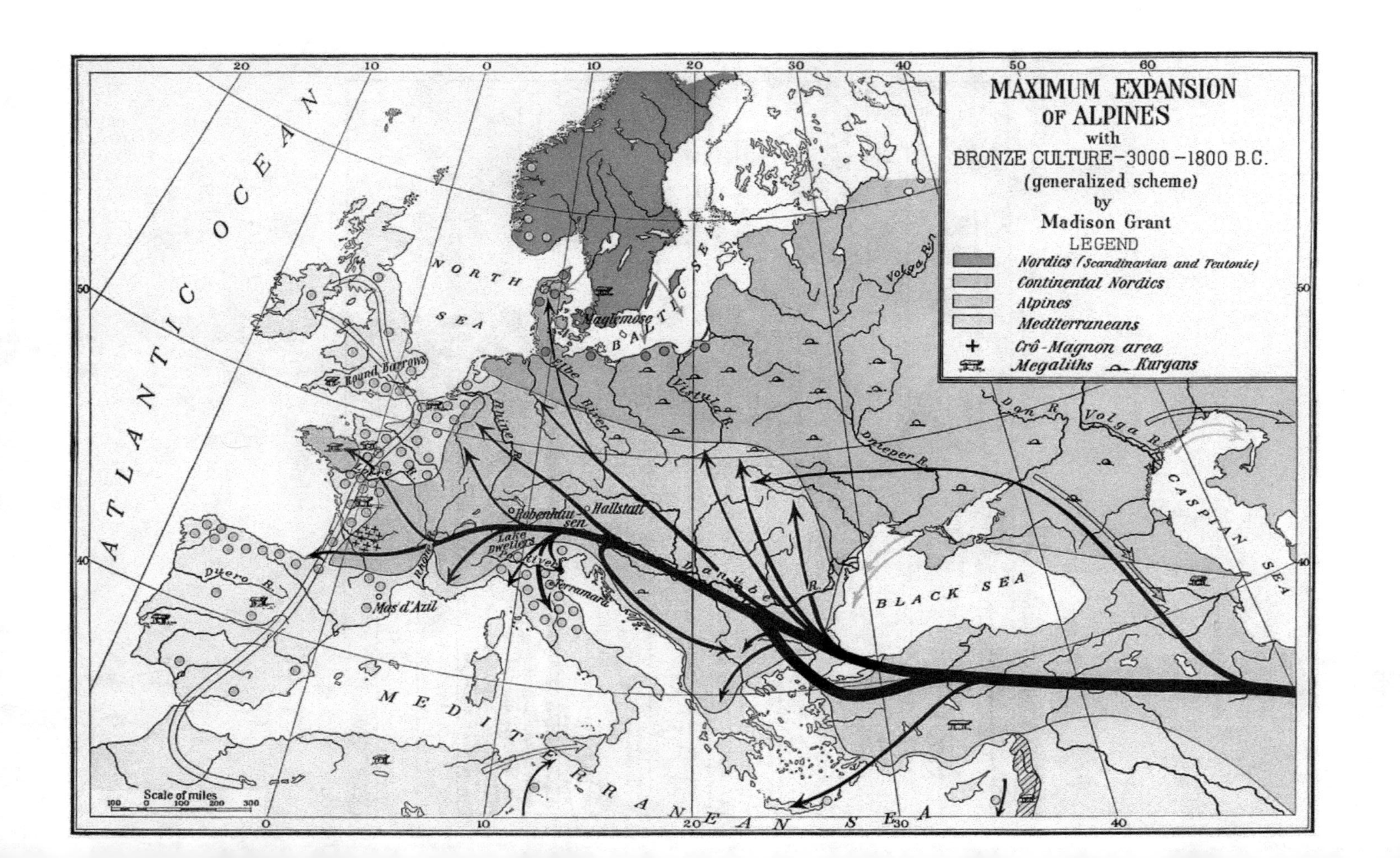

MAXIMUM EXPANSION
OF ALPINES
with
BRONZE CULTURE–3000–1800 B.C.
(generalized scheme)
by
Madison Grant
LEGEND
Nordics (Scandinavian and Teutonic)
Continental Nordics
Alpines
Mediterraneans
Crô-Magnon area
Megaliths
Kurgans
ATLANTIC OCEAN
NORTH SEA
BALTIC SEA
BLACK SEA
CASPIAN SEA
MEDITERRANEAN SEA
Maglemose
Round Barrows
Elbe River
Vistula R.
Volga R.
Don R.
Dnieper R.
Rhine R.
Loire R.
Rhone
Duero R.
Danube R.
Po River
Hallstatt
Robenhausen
Lake Dwellers
Terramara
Mas d'Azil
Scale of miles
100 0 100 200 300

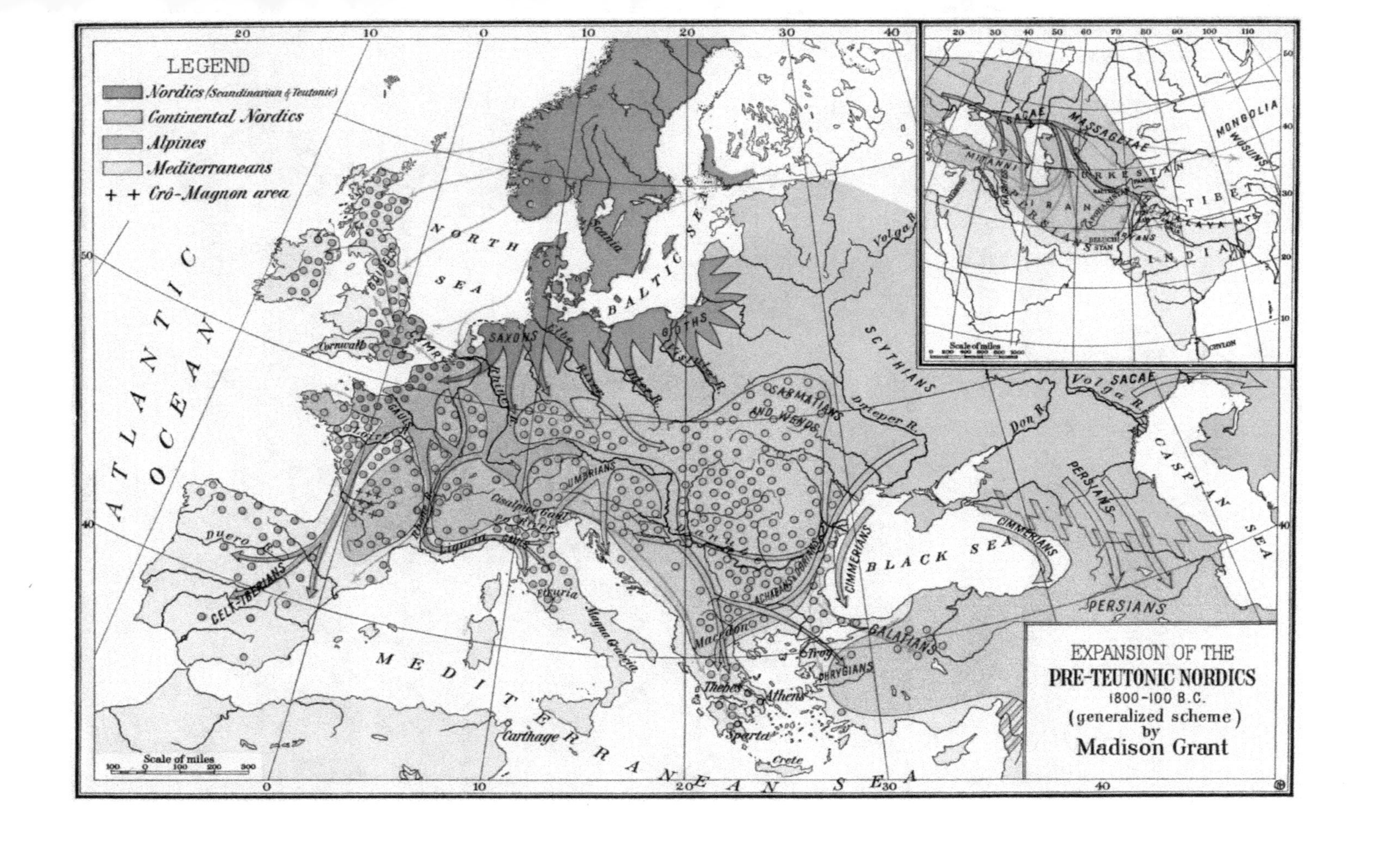
EXPANSION OF THE
PRE-TEUTONIC NORDICS
1800-100 B.C.
(generalized scheme)
by
Madison Grant
LEGEND
Nordics (Scandinavian & Teutonic)
Continental Nordics
Alpines
Mediterraneans
+ + Crô-Magnon area
ATLANTIC OCEAN
NORTH SEA
BALTIC SEA
BLACK SEA
CASPIAN SEA
MEDITERRANEAN SEA
SAXONS
GOTHS
SCYTHIANS
SARMATIANS AND WENDS
CIMMERIANS
PERSIANS
GALATIANS
PHRYGIANS
UMBRIANS
GAULS
CELT-IBERIANS
SACAE
Scania
Cornwall
Elbe River
Oder R.
Vistula R.
Dnieper R.
Don R.
Volga R.
Duero
Cisalpine Gaul
Etruria
Magna Graecia
Carthage
Macedon
Thebes
Athens
Sparta
Crete
Troy
Scale of miles
MASSAGETAE
MONGOLIA
WUSUNS
MITANNI
TURKESTAN
TIBET
IRAN
INDIA
ARYANS
CEYLON

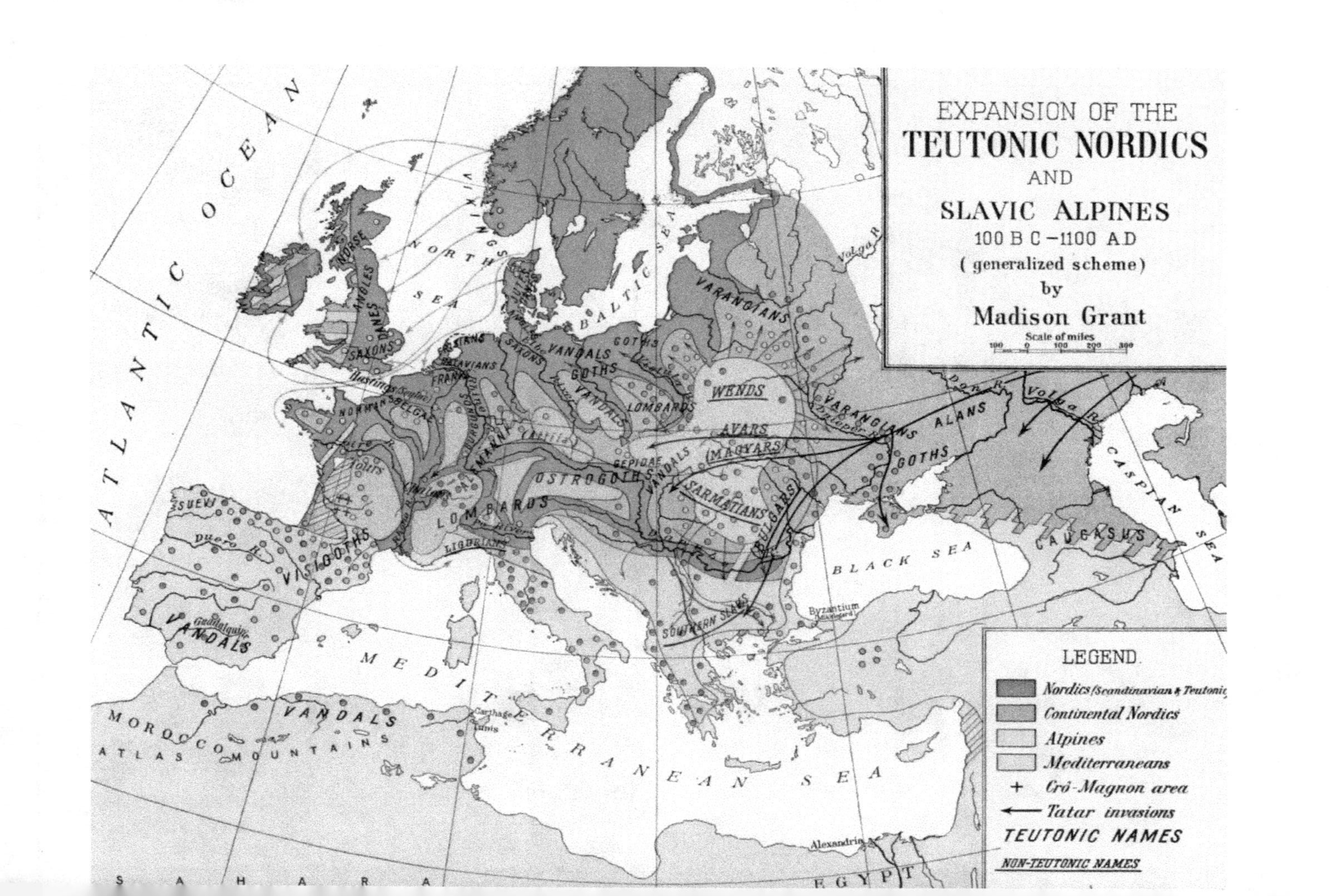
EXPANSION OF THE
TEUTONIC NORDICS
AND
SLAVIC ALPINES
100 B C – 1100 A D
( generalized scheme)
by
Madison Grant
Scale of miles
100 0 100 200 300
LEGEND
Nordics (Scandinavian & Teutonic)
Continental Nordics
Alpines
Mediterraneans
Crô-Magnon area
Tatar invasions
TEUTONIC NAMES
NON-TEUTONIC NAMES
ATLANTIC OCEAN
NORTH SEA
BALTIC SEA
BLACK SEA
CASPIAN SEA
MEDITERRANEAN SEA
VIKINGS
NORSE
ANGLES
DANES
SAXONS
VANDALS
GOTHS
VARANGIANS
WENDS
LOMBARDS
AVARS
MAGYARS
GEPIDAE
SARMATIANS
BULGARS
ALANS
OSTROGOTHS
VISIGOTHS
SUEVI
LIGURIANS
SOUTHERN SLAVS
MOROCCO
ATLAS MOUNTAINS
SAHARA
EGYPT
CAUCASUS
Byzantium
Carthage
Tunis
Alexandria
Volga R.
Don R.
Duero R.
Tours

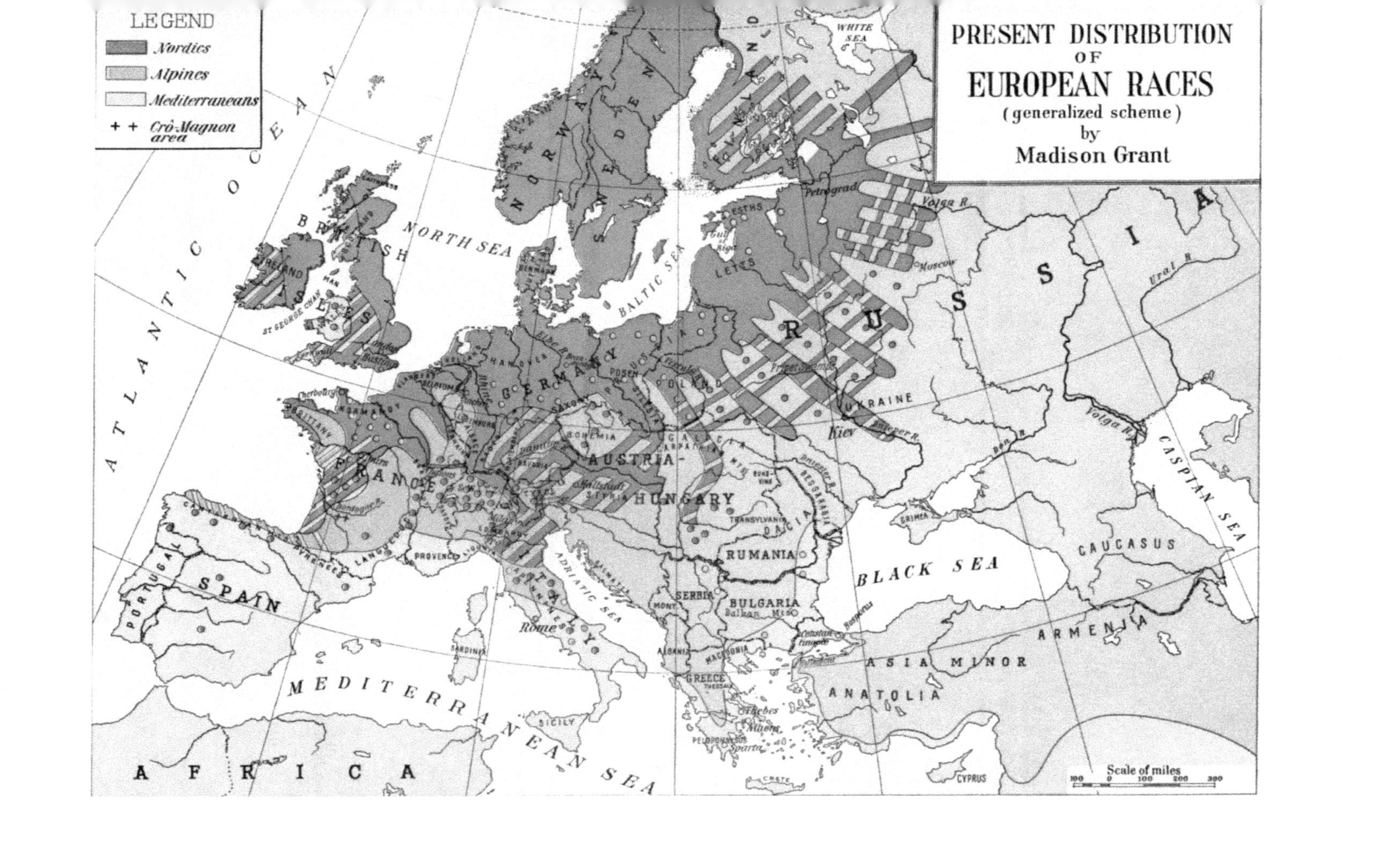
PRESENT DISTRIBUTION
OF
EUROPEAN RACES
(generalized scheme)
by
Madison Grant
LEGEND
Nordics
Alpines
Mediterraneans
+ + Crô-Magnon area
ATLANTIC OCEAN
NORTH SEA
BRITISH ISLES
IRELAND
NORWAY
SWEDEN
FINLAND
WHITE SEA
BALTIC SEA
Petrograd
Volga R.
Moscow
ESTHS
LETTS
Gulf of Riga
RUSSIA
ASIA
Ural R.
CASPIAN SEA
CAUCASUS
UKRAINE
Kiev
Dnieper R.
Don R.
CRIMEA
BLACK SEA
GERMANY
HANOVER
POSEN
POLAND
Rhine
SAXONY
BOHEMIA
AUSTRIA-HUNGARY
STYRIA
TRANSYLVANIA
DACIA
BESSARABIA
RUMANIA
SERBIA
BULGARIA
Balkan Mts.
MONT.
ALBANIA
MACEDONIA
GREECE
THESSALY
Thebes
Athens
PELOPONNESUS
Sparta
CRETE
Bosporus
ASIA MINOR
ANATOLIA
ARMENIA
CYPRUS
FRANCE
NORMANDY
BRITTANY
Cherbourg
PROVENCE
PYRENEES
SPAIN
PORTUGAL
ITALY
Rome
SARDINIA
SICILY
ADRIATIC SEA
MEDITERRANEAN SEA
AFRICA
Scale of miles
100 0 100 200 300

# NOTES

# NOTES

# NOTES

# NOTES

# NOTES

CPSIA information can be obtained
at www.ICGtesting.com
Printed in the USA
BVHW091031101220
595268BV00002B/7